Advance Praise

The author is one of India's most renowned journalists and a former member of the Bruntland Commission. He has been at the heart of India's national discussion of sustainable development for decades and brings to the subject a wide international experience, a fierce understanding of technology and science, and a gift for communicating complex subjects in accessible and engaging language.

Since the brief surge of optimism that followed the signing of Paris Agreement a starker sense of reality has set in. The pledges agreed in Paris take the world to an increase of 3.5°C, well beyond any idea of safe levels. Already there is alarming evidence of an accelerating warming that risks pushing the world beyond the threshold at which human action could prevent catastrophe. The Trump administration is spearheading a pushback by fossil fuel interests that will slow the effort further.

Against this background, Prem Shankar Jha offers a clear way forward. The book is an exploration of the options for avoiding dangerous climate change, starting from the essential question: Given that we know how to effect a transition to a low-carbon economy, and given the catastrophic risk to human civilisations that climate change represents, why have we failed to respond?

—Isabel Hilton
Columnist and reviewer for *The Guardian*

This is an extremely timely book. Today, as the perception of the risks facing humanity from the worsening impact of climate change is beginning to sink into the public's consciousness, this book comes as an antidote to despair. For it shows not only that the technologies needed for an immediate shift out of fossil fuels have already been found, but that some of them have been around for almost a century. Perversely, on account of inertia in the system and the power of vested interests, these have been neglected to the detriment of all species on planet earth, including human society. Prem Shankar Jha's book also solves a riddle that has puzzled the climate science community for more than a decade: when the world's governments could create the Intergovernmental Panel on Climate Change (IPCC) to identify and assess the threat of climate change, why did they fail utterly to convert into action the solutions identified by the IPCC?

—**R.K. Pachauri**
Former Chairman of the Intergovernmental Panel on
Climate Change

Dawn of the

SOLAR AGE

Dawn of the

SOLAR AGE

An End to Global Warming and to Fear

PREM SHANKAR JHA

Los Angeles | London | New Delhi
Singapore | Washington DC | Melbourne

First published in 2018 by

 SAGE Publications India Pvt Ltd
B1/I-1 Mohan Cooperative Industrial Area
Mathura Road, New Delhi 110 044, India
www.sagepub.in

SAGE Publications Inc
2455 Teller Road
Thousand Oaks, California 91320, USA

SAGE Publications Ltd
1 Oliver's Yard, 55 City Road
London EC1Y 1SP, United Kingdom

SAGE Publications Asia-Pacific Pte Ltd
3 Church Street
#10-04 Samsung Hub
Singapore 049483

Published by Vivek Mehra for SAGE Publications India Pvt Ltd, typeset in 11.5/14 pt Adobe Garamond by Zaza Eunice, Hosur, Tamil Nadu, India and printed at Saurabh Printers Pvt Ltd, Greater Noida.

Library of Congress Cataloging-in-Publication Data

Name: Jha, Prem Shankar, author.
Title: Dawn of the solar age : an end to global warming and to fear / Prem
  Shankar Jha.
Description: New Delhi, India : SAGE Publications India Pvt., Ltd., 2018. |
  Includes bibliographical references and index.
Identifiers: LCCN 2017040816| ISBN 9789386602992 (print pb) | ISBN
  9789352800049 (e pub 2.0) | ISBN 9789352800032 (e book)
Subjects: LCSH: Solar energy. | Global warming. | Atmospheric carbon dioxide. |
  Renewable energy sources.
Classification: LCC TJ810 .J53 2018 | DDC 333.792/3—dc23 LC record available at https://lccn.loc.
gov/2017040816

ISBN: 978-93-866-0299-2 (PB)

SAGE Team: Rajesh Dey, Alekha Chandra Jena, Madhurima Thapa and Rajinder Kaur

To my grandchildren, Jai and Amaya, and their children whom I will not see, with an apology for the terrible world my generation is leaving to them, and the hope that they will remember that I tried to make it a little better.

Thank you for choosing a SAGE product!
If you have any comment, observation or feedback,
I would like to personally hear from you.

Please write to me at **contactceo@sagepub.in**

Vivek Mehra, Managing Director and CEO, SAGE India.

Bulk Sales

SAGE India offers special discounts
for purchase of books in bulk.
We also make available special imprints
and excerpts from our books on demand.

For orders and enquiries, write to us at

Marketing Department
SAGE Publications India Pvt Ltd
B1/I-1, Mohan Cooperative Industrial Area
Mathura Road, Post Bag 7
New Delhi 110044, India

E-mail us at **marketing@sagepub.in**

Get to know more about SAGE

Be invited to SAGE events, get on our mailing list.
Write today to **marketing@sagepub.in**

This book is also available as an e-book.

Contents

Foreword

In 1982, when I was the Acting Deputy Chairman of the Planning Commission of India, Prem Shankar Jha brought to the Commission a proposal for converting municipal solid waste into transport fuels. This could be done, he explained, through the destructive distillation of urban solid waste to yield carbon monoxide and hydrogen, which could then be recombined, using the right catalysts, temperatures and pressures, into methanol and other transport fuels. The technology was well understood and had been used to produce methanol from wood or coal since the Second World War. It could, in fact, convert any form of biomass into transport fuels.

Harnessing this technology could, therefore, greatly reduce pollution and disease in the cities, provide the stimulus India needed to increase the land area under tree cover, create millions of jobs in sylviculture and agroforestry, and double farm incomes if not more by giving crop residues a commercial value equal to, if not greater than, the value of the principal crop. It would also save scarce foreign exchange.

The Commission accepted the proposal and notified the concerned ministries in the central and state governments of India, but the crash of oil prices that followed three years later ensured that the project remained stillborn.

Today, when the threat of climate change has become very real, I am greatly pleased to find that Jha has not only continued to follow the research done in laboratories all over the world but has also brought the results together in a book to show how global warming can be limited not just to the 2°C that governments hope will keep the world safe, but the 1.5°C that climate scientists believe is all that the world can bear.

Jha has not stopped there but gone on to show how the public awareness of these technologies has been suppressed to make it continue believing that there are no alternatives to fossil fuels. Of still greater value are his insights into why powerful governments, large global corporations and even a section of its scientists have collaborated in this deception.

The answer, according to him, does not lie in their narrow pursuit of self-interest, but in a profound reluctance among those who control the levers of power today, to face the political consequences of a shift in the energy base of society from coal, oil and gas, to solar and biomass energy.

It lies in the fact that just as the concentration of coal in the geographical North gave rise to the hegemony of Europe and North America, a shift in the energy base to the sun and biomass could have the opposite effect. Political scientists will find his insights into the myriad ways in which ideology and information have been harnessed to slow this down the most fascinating, if not the least controversial, part of the book.

M.S. Swaminathan
Founder Chairman, M.S. Swaminathan Research
Foundation
Former Director General of the International
Rice Research Institute

Preface

Once a wraith shrouded in the mists of the future, climate change has emerged as a threat to humanity that is too disturbing to contemplate. Beginning with the first reports of the World Meteorological Organization in the late 1960s, every succeeding report has said essentially the same thing: that the world is growing warmer at a historically unprecedented rate, and that an accumulation of carbon dioxide, and other greenhouse gases released by the burning of fossil fuels, is the main cause. Today, with each of the last 14 years having been hotter than its predecessor, and unprecedented weather events having become routine, the room to doubt that it has begun, and may even be accelerating, has shrunk till there is almost none left.

My purpose of writing this book is not to add one more voice to the rising clamour of alarm. It is to show that while the threat is very real, *we already have the capacity not only to arrest it, but to create, in the process, a cleaner, more egalitarian and, above all, a more peaceful world.* The key to doing this lies in shifting our energy base out of fossil fuels within the next half-century at most. If we do this, we will not only limit the climate change to a level that human beings and most other living species will be able to adjust to, but also eliminate most of the causes of the conflict that is racking the world. Which of these two legacies we will leave for our grandchildren will depend upon the choices we make in the next decade. Tarry longer, and the choice will be made for us.

As the huge circulation of an article published by *New York* magazine, on 10 July 2017,[1] showed, this sense of urgency has begun to creep into the public's awareness. But with no credible solutions on offer it has been bitterly criticised on the ground that it can only sow panic. The desire to avoid this is also the reason why not only the governments of the largest countries of the world but also the scientific community whose consensus views have been presented to the world by the Intergovernmental Panel on Climate Change have been extremely conservative in their projections of the likely impact of global warming, so far.

Another, more deep-seated, reason for our conservatism is that human beings are biologically conditioned to react only to immediate and tangible threats. Even with the acceleration that seems to have set in, climate change does not fit this bill. But there is a third, so far barely acknowledged, reason for our reluctance: The very same 'civilisation' that has bestowed unimaginable affluence upon humanity, and so greatly widened the frontiers of knowledge that we now have a pretty accurate idea of what the universe looked like a mere 300,000 years after it was born,[2] has also enslaved our minds and made us blind to the dangers it is now spawning.

Its moving force today is the market economy. The competition inherent in it is responsible for the dizzying acceleration of science and technology that has brought the world where it is today. In the past, human beings thanked God for 'rest and home and all things good', and looked to Him to save us from misfortune. Today we have transferred most of that faith to the market. This has robbed us of the capacity to control the market even when the way it works is patently driving humanity ever closer to suicide.

The proof of this is to be found in the vast gap between what we profess and what we actually do. Look at the Manhattan skyline from the New Jersey Shore of the Hudson, or the London skyline from the Post Office Tower, and ask yourself whether the night sky is any less brightly lit now than it was 40 years ago, and you will find the gap. 'But these are fluorescent, Halogen, and LED lights,' you might say. 'They consume a tenth of the power that tungsten bulbs consumed'. And you might then add, for good measure, 'I am already running a Prius, and plan to switch to a Tesla the next time round. Am I not doing my bit to prevent Global warming?' You are, but there are 10 times as many high-rise buildings in

Manhattan as there were 30 years ago. Similarly, hybrids only halve fuel consumption at the very best, and the power for the electric car battery still comes from a power station that burns coal or gas. The grim truth is that so long as the population keeps growing, incomes keep rising and people continue to aspire to a better life, using fossil fuels more efficiently will only postpone the fifth mass annihilation of species that we are in the midst of. Only this time we will be among the victims.

Climate change is not the only threat that humanity faces. A more immediate one is that fossil fuels are getting harder to find all the time. The struggle to do so and to control their use is therefore intensifying and taking the world closer to war. The 'point of inflexion' where the discovery of new, exploitable, reserves of fossil fuels starts falling behind the growth of consumption is on the point of being reached in the case of oil, and is expected to be reached before the end of this century for coal. If we are still dependent then upon coal, gas and oil for most of our commercial energy, the anxiety that the major powers of the world are beginning to feel will turn into paranoia.

The threat posed to humanity by the approaching exhaustion of fossil fuels was first spelt out by the Club of Rome in 1972. Its report, 'The Limits to Growth', had warned that if population, income and consumption continued to grow at the exponential rates recorded in the previous quarter of a century, the global economic system would collapse around the middle of the 21st century in famine, pestilence and war. In the 1970s, the report was dismissed as being alarmist, but a recent study has shown that in 2002 the world was precisely on the track it had predicted 30 years earlier.

The outliers of the approaching storm are already overhead. Even as warming oceans, prolonged droughts and flash floods are endangering the food supply chain in arid and semi-arid regions, the race among the major powers to corner the mineral resources that remain is plunging an expanding area of the world into a state of near-permanent war. War and extreme civil conflict have engulfed Iraq, Libya, Syria, Egypt, Yemen, Saudi Arabia, Qatar and Bahrein, and is lapping at the doorsteps of Iran. Not coincidentally, all but two of these are oil- and gas-producing states. Somalia and Sudan are failed or failing states.

War and famine are combining to trigger mass migrations and where this is resisted, as in Darfur, to genocide. The fear of hordes of starving African and Middle Eastern refugees descending upon their countries is already changing the political climate in Europe and the United States. If desperation and paranoia continue to mount, it is perfectly possible that by the middle of the 21st century the use of nuclear weapons may no longer be unthinkable.

Shifting the energy base of the global economy from fossil fuels to solar energy in one or more of its myriad forms will avert both the threats outlined previously. Solar power is clean; it is fairly evenly distributed across the inhabited world, and it is limitless. But it is here that we encounter a singular anomaly. The technologies that can do this are now well known. What is more, they have been known to scientists for four decades in the case of solar electricity and almost a hundred years in the case of transport fuels. So why does the general public continue to believe that fossil fuels are indispensable?

Asking this question opens a Pandora's box of doubts. Could it be because human memory is short and incapable of retaining the deluge of information that it is now subjected to? Could it be because the alternatives to fossil fuels are untested and considered to be too expensive and too limited in supply to be taken seriously? Could the knowledge have been deliberately suppressed by the energy lobby? All three explanations touch only the fringes of truth.

The answer lies in a far older truth: from the taming of fire to the splitting of the atom, civilisation has been built upon mankind's capacity to harness and control various sources of energy. The way in which it has done so has shaped the structure of every civilisation since the last ice age: It explains the wealth and pre-eminence of Greece, Persia, China and India two thousand years ago; of the Arabs Caliphates and Byzantium a millennium ago and of Europe and North America in the last five hundred years.

Today, nearly every policy maker (except Donald Trump) concedes that another energy shift has become imperative, but no one fully understands how it will affect the power balance between people within and between nations. All that we know for certain is that such a fundamental shift is bound to create vast numbers of new winners and losers, and the biggest losers would be those who have invested most heavily in the existing economic and political power structure of the world. So as the need to shift becomes more urgent, the effort to first discredit the science that can make it possible, then to capture emerging technologies in order to control the pace of their introduction, and finally to suppress, or at least discredit the emerging technologies that have the greatest capacity to upset current power relations in the world, has steadily gathered momentum.

This book therefore has three goals: to explode the myth that there is no way to end our dependence upon fossil fuels; to under-stand how and why the world has been persuaded to believe so and to catch a glimpse of another world that we can leave behind for our grandchildren if we make the shift out of fossil fuels as rapidly as technology and economics have now made possible. Suffice it to say that it will be very different from the stormy, war-racked world that we live in today.

I first glimpsed this world in 1980 when S. Paul, a scientist friend in New Delhi, walked into my office and asked me whether my newspaper, the *Financial Express*, would publish an article on a way out of the oil price crunch that was threatening to strangle the Indian economy. When I asked him what he had in mind he said:

Oil can be obtained by burning any biomass in a limited supply of air. Various tarry oils have been by-products of the making of charcoal for hundreds of years. The technology has now been developed to the point where it can yield almost any bio-fuel.

In the 1970s, India depended upon imported oil to meet 70 per cent of its needs, so I had watched Brazil's recent success in substituting gasoline with ethanol with envy and wondered whether India could do something similar. When I asked Paul whether this was being done anywhere, he drew my attention to a White Paper published by the New Zealand government two years earlier that had drawn up a plan to gasify a part of the country's annual harvest of *pinus radiata* in order to produce methanol. The methanol could be used as an octane booster, as Brazil was doing with ethanol, or be converted into gasoline by a simple process that had been developed by Mobil and is in use till this day.

The New Zealand White Paper lit a fuse in me that is still burning today, for not only was methanol an older, tried-and-tested transport fuel than ethanol, but it could be produced not only from wood but from any kind of biomass, be it wood shavings and timber residues, to agricultural crop residues or urban solid waste. It did not, therefore, suffer from the supply constraints of ethanol. As I dug deeper into the subject, I found myself looking at India's filthy cities, its ravaged landscape, its poverty-stricken villages and its toiling, emaciated peasants, with different eyes. If garbage could become the feedstock for transport fuels, it would cease to be a nuisance and acquire immense value. In a matter of months, therefore, all the accumulated waste lying in and around urban garbage dumps, and along hundreds of miles of roads and rail tracks, would simply disappear, and re-emerge as diesel, gasoline or aviation turbine fuel.

The past half-century had seen the slaughter of India's forests and decimation of its once abundant wildlife on a scale close to what Brazil and Malaysia are experiencing now. The New Zealand plan offered us a chance to re-afforest the 100 million hectares of forest lands that had been denuded since India became independent. This would take the pressure off, and save, the natural forests and wildlife that remained. Reforestation would also provide employment to millions upon millions of youth in the villages who were languishing for want of employment. The conversion of what are now useless, or low-value, crop residues into transport fuels would give these residues a market value that would double farmers' incomes at the very least. Best of all, it would rejoin the umbilical cord between agriculture and industry that had been severed when humanity turned from wind and water power to fossil fuels to quench its thirst for energy.

So at age 45, full of excitement and optimism, I took a proposal to gasify urban solid waste to produce power at the first stage, and methanol at the second, to the Planning Commission of India and a few months later to Prime Minister Indira Gandhi. Mrs Gandhi immediately agreed that this was a technology that needed to be harnessed and forwarded the proposal to the Government of Maharashtra. The planning commission also forwarded the proposal to other state governments for appraisal and implementation. Ten days after seeing Mrs Gandhi I found myself the rapporteur of an Indian delegation sent around the world to appraise available technologies for converting urban solid waste into fuel gas (also known as synthesis gas).

Then in the next few months I got my first taste of the real world, for nothing happened. Two years later, oil prices collapsed. Since then the proposal to gasify municipal solid waste to produce transport fuels has gathered dust. Today very few outside India's chemistry laboratories even know that such a technology exists.

It has taken me more than 30 years to understand that this torpor was not a peculiarly Indian affliction and to identify its cause. Before the Copenhagen Climate Summit of 2009, Western leaders, from the former British Prime Minister Tony Blair to Danish Prime minister Anders Fogh Rasmussen, kept insisting that the technologies that could usher in a 'low carbon' world had been identified and needed only a little political will to be harnessed. But in the 15 meetings of the signatories to the UN Framework Convention on Climate Change (UNFCCC) that culminated in the Copenhagen Climate Summit of 2009, and the 6 meetings, culminating in the Paris summit, that have followed, no one has spoken a single word about the technologies that can make the shift possible and, more importantly, those that cannot. Inevitably, therefore, there has not been even an hour of discussion of the cost of energy delivered by these supposedly well-known technologies and of the yardstick (in terms of future fossil energy costs) against which these should be measured. This book suggests that the explanation of this anomaly is to be found in the way that market economies function and even more fundamentally, in our enslavement to an idealised concept of 'the market', which has prevented us from understanding the way in which it actually functions. Much of this book is devoted to explaining this anomaly.

Notes and references

[1]   David Wallace-Wells, 'Our Uninhabitable Earth'. *New York,* 10 July 2017.

[2]   https://naturalhistory.si.edu/exhibits/evolving-universe/posters/4.0.1.pdf

Acknowledgements

From a simple search for alternative feedstock to oil for the production of transport fuels that began in 1980, to a quite recent realisation that it is not the lack of alternatives to fossil fuels, but the increasingly slavish relationship of governments to the dictates of the global 'free' market, which has brought us to the tipping point in climate change, the insights I have tried to present in this book has taken me almost 40 years to develop. During my long journey, I have been helped by innumerable people who have taken the time and trouble to read, comment on and correct what I have written and encouraged me to take the next steps in my quest. But I will do my best to remember them and thank them for their guidance.

They include two great editors of the *Times of India*, Sham Lal Jain and Girilal Jain, who gave me the freedom to wander where my curiosity and concerns took me in my weekly columns on the economy in the newspaper. I would like to thank Dr M.S. Swaminathan, former Director General of the International Rice Research Institute (IRRI), who not only invited me to present my ideas to the Planning Commission of India in 1982 but also invited me to do so again at a conference on whole crop utilisation at IRRI in the Philippines the next year. I would like to thank former Chairman of the Intergovernmental Panel on Climate Change Dr R.K. Pachauri for allowing me to present my ideas at conferences of the Tata Energy Research Institute (now The Energy and Resources Institute; TERI) as far separated as in 1983 and 2015, and for bringing my writings on energy conservation and renewable energy to the attention of the International Institute of Energy Economics in Washington in 1986. I am, in particular, most grateful to Dr Ashok Khosla, former Co-President of the Club of Rome and President of the International Union for the Conservation of Nature, for recommending me for membership of the energy panel of the World Commission on Environment and Development in 1985. Had that not happened I might never have made the transition from worrying about foreign exchange shortages and air and water pollution, to worrying about climate change.

From writing a book to getting it published is a long and difficult road, especially when there are already 21,000 books on global warming and climate change in the market. So I am profoundly grateful to my agent Laura Susijn for undertaking the daunting task of persuading editors that this book has something important to say. Her task was not easy, but the peer review process proved invaluable to me, for it exposed me to the concerns and, in some cases, prejudices that my book would encounter and that I had therefore to guard against. This process of reflection led to insights that have transformed the original draft. I have, therefore, to thank all the peer reviewers who took the trouble to read my book and pen their comments to thank, and I would like to do so even though I will never know who they are. I would like to thank the editor at SAGE, New Delhi, who not only read my manuscript with painstaking attention, but convinced the management that the book, although written for the general reader, had sufficient scholastic merit to fit in with SAGE's image as an academic publisher.

Part I

Where We Are Today

Chapter One
Elements of a Perfect Storm

Describes the growth of our understanding of climate change, from the writings of James Lovelock in the 1970s, till the findings of scientists in the decade after Al Gore and the IPCC's fourth assessment report put the world on red alert. Explains the 'tipping point'.

Climatologists have been warning us for more than two decades that the first sign of accelerating climate change will be an increase in the frequency of extreme weather events all over the world. Heat and cold waves will become more intense, droughts, floods, snowstorms and cloudbursts more frequent and more destructive. By that yardstick, it has already begun. On 22 December 2016, the temperature at the North Pole was +0.4°C when it should have been −25°C if not lower. For the entire month before that, the polar region had been in the grip of what was for it a 'heatwave', for on 23 November the temperature at the North Pole was only −5°C.[1] The cause was the disruption of the polar vortex, a system of strong winds in the stratosphere that keeps the frigid air above the North Pole locked there. The disturbance of the vortex has resulted from warm air flowing north. This has displaced the envelope of high winds—the jet stream—that contains the frigid Arctic air over the North Pole and pushed it into a cold air ridge across Siberia, Canada and Northern China. A similar shift of the polar vortex had caused the Hudson to freeze at New York in February 2014.[2]

The last six years, 2011 till 2016, have been the hottest the planet has experienced since records began to be kept. What is still more alarming, the global temperature anomaly, that is the amount by which the world's average global air temperature exceeds what it used to be in the pre-industrial age, has not only been rising steadily, but has virtually become a straight line. This anomaly was fairly steady at 0.2°C till 1960, but had then risen rapidly to 0.8°C in 2010. Since then it has been climbing exponentially and has reached 1.09°C in 2016.[3]

Not surprisingly, 2016 had been the hottest year in human memory since records began to be kept. The preceding summer had been the hottest Europe had ever known. In the winter of 2014, Boston had received a hundred inches of snow in a single storm. In the summer of 2013, the maximum temperature in New York had risen for the second time in three years to 104°F. In October 2012, the city, and much of the east coast of the United States, was devastated by tropical storm Sandy which was able to attain its monstrous power because the Gulf Stream was running 5°C warmer than normal, generating unprecedented amounts of evaporation.

In the same year, unprecedented floods in January in Brazil were followed by the worst drought in 50 years in the north-east part of the country, while 50 major wildfires raged across Chile. In 2013, Europe experienced its coldest winter in a quarter of a century while across the Atlantic, the United States had its mildest winter on record. In the following summer, 57 per cent of the United States experienced the worst drought in its recorded history, and wildfires in 13 states destroyed 1.36 million acres of forests. At the same time, parts of China were being deluged by torrential rains that required five million people to be evacuated. Record-breaking floods also occurred in Queensland, Australia, side by side with a devastating drought that put 10 million lives in jeopardy in the Sahel region of Africa.[4]

In India, the frequency of extreme weather events has increased sharply in the past decade. The heaviest deluge in Mumbai's recorded history dumped 94 cm (more than 3 feet) of rain on the northern parts of the city in less than a day causing heavy loss of life and property. In 2010, the cold, trans-Himalayan desert region of Ladakh, where people have been building their homes with unbaked bricks since times immemorial, received 10 inches of rain in a single downpour. This literally washed away most of Leh, the capital of Ladakh, and scores of other villages.

In 2013, an unprecedented cloudburst in the central Himalayas in India swept away the entire pilgrimage town of Kedarnath, and an entire hydro-power station on the Ganges river, killing between 5,000 and 10,000 pilgrims and inhabitants of the area. The years 2015 and 2016 saw two of the worst droughts that southern part of India has ever experienced. In the same two years, unprecedented cloudbursts submerged almost the whole of the southern port city of Chennai on the Bay of Bengal.[5]

In 2017, India faced a heatwave at a time when it had never experienced one before in living memory. This was in March, when normally, spring is at its height, and a hundred varieties of trees and shrubs burst into bloom in northern and central India before their long summer hibernation. On 28 March, the town of Bhira in Maharashtra recorded a temperature of 46.5°C, making it the second hottest place in the world. On the same day, the Indian Meteorological Department (IMD) and the National Disaster Management Authority issued a warning to seven states of India covering almost half of the country that they would experience a heatwave for the next four days and should take the necessary precautions to prevent deaths from heatstroke and thirst.[6] The IMD defines heatwave as a condition in which the mean temperature rises by 5°–6°C above normal. Throughout the last week of March, the mean temperature in Delhi remained 7°C above normal.[7]

Human beings are the culprits

Human action is directly responsible for most, if not all, of these events. Figure 1.1 reveals the close correlation between the

Figure 1.1: CO_2 concentration in the atmosphere and mean air temperature in AD 1000–2000[9]

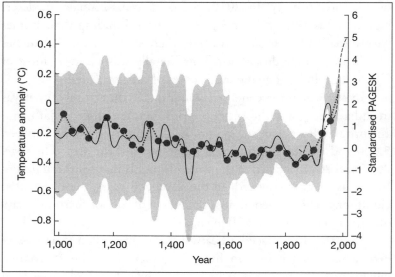

Source: See note 9.

concentration of CO_2 in the atmosphere and the mean global air temperature. It shows that after remaining stable at around 280 parts per million (ppm) for 800 years, the CO_2 content of the air has risen very sharply since the beginning of the 20th century. As it has increased, it has trapped more and more of the heat that the earth radiates out at night. By 2010, this had created an imbalance of approximately 0.5 W/m² per day, equivalent to the energy released by 400,000 Hiroshima-sized nuclear bombs.[8] This imbalance cannot go on increasing forever, so the only way the earth can adjust is to force more heat out at night through the denser blanket of greenhouse gases (GHGs). To do that it 'has had to grow warmer'.

Since 2010, the rise in temperature has accelerated dramatically. Although the very strong El Niño from 2014 to mid-2016 has undoubtedly contributed to it, its prime cause remains the accumulation of CO_2 and other GHGs. The Global Carbon Project (GCP), an international collaboration of 49 scientists, reported that between 1990 and 2013, CO_2 emissions had risen by an astonishing 61 per cent.[10] They estimated that if the rise continued unabated, the mean air temperature would rise by more than 3° by 2100 AD. And that would only be the beginning, because the warming would continue for hundreds of years.

Much of the CO_2 that will get us to this dismal situation is already in the pipeline. The United Nations has calculated that to have a reasonable chance of meeting the 2°C target, the world has to limit CO_2 emissions to 1,000 giga (billion) tonnes (Gt) by 2100. But even if the Paris commitments are met—and that is quite a big if—the world will still have added around 723 Gt of CO_2 by 2030. To stay within the 1,000 Gt limit, therefore, it will have to bring down annual CO_2 emissions from 40 Gt a year to 0 in the subsequent 14 years. Without a rapid exit from fossil fuels well before the end of this century, and a subsequent removal of CO_2 from the atmosphere, this is manifestly impossible.[11]

But the entire developing world is headed in the opposite direction. And regardless of what they profess, so are the governments of the 20 major economies. The pressures of domestic and international politics, and the enormous influence of the energy lobby and its associated industries, have so completely distorted public perceptions that most people now consider 'fracking', which has been vigorously embraced by the governments of the United States, China, Israel, the United Kingdom and western Europe, as a way of creating a 'low carbon economy' because it will hasten the replacement of coal with natural gas in their thermal power plants. James Hansen's derisive comment that 'these governments seem to have forgotten that when there is already too much CO_2 in the atmosphere, slowing down its further accumulation will at most postpone the day of reckoning, not avert it' has fallen on deaf ears.[12]

To a layman, a temperature rise of even 1.09°C may only seem an inconvenience, not a threat. But this average is deceptive.

Global warming is unevenly distributed and is greatest in the part of the planet where it can do the most to destabilise the climate. This is the Arctic zone of the northern hemisphere, which stretches from Alaska through Canada and Russia to China and the Bering Sea. In these regions, as the Intergovernmental Panel on Climate Change (IPCC) pointed out in its fourth assessment report in 2007, the rise in air temperature has consistently been two to three times greater than the global average. But this ratio too seems to be changing for, in the past six years, against the 1.09°C average, the rise of temperature in some of the Arctic regions of Russia has been 6° to 7° above normal, while in other parts of Russia, Canada and Alaska, it has been 3° to 4° above the average.[13] The most disturbing temperature anomaly occurred, however, at the North Pole which experienced a 'heatwave', with temperatures skirting the melting point all through November and December, while 12 cities in Russian Siberia experienced temperatures 40°C below normal temperatures from as early a November 15 (Figure 1.2).[14]

In a report that it released at the Climate Change Conference in Marrakesh in November 2016, the World Meteorological Organization estimated that more than half of the major extreme weather events from 2011 to 2015 could be traced back to human-induced global warming. These had cost 300,000 human lives, mostly because of drought. The unprecedented temperatures had also warmed the seas, increased evaporation and loaded the warmer air with moisture. Like hurricane Sandy in 2012, this had enhanced the destructive power of a dozen hurricanes and tropical storms that followed in its wake.[15]

Five years later Tropical Storm Harvey dumped an estimated 51 inches of rain on Houston, the highest single precipitation recorded in the United States.

Scepticism still abounds

But scepticism still abounds and, to make matters worse, one of the most trenchant, but ill-informed among the sceptics is now

Figure I.2: Global temperatures—Change from pre-industrial levels

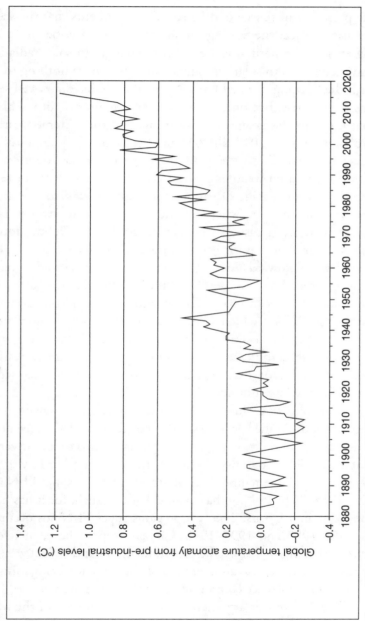

Global temperature anomaly from pre-industrial levels (°C)

Source: NOAA, NASA, UK Met Office/CRU.

the president of the United States. The arguments that the sceptics use to reject the findings of climate science ascribe all warming episodes to natural causes such as changes in solar radiation
and oscillations in solar and lunar orbits that have nothing to do
with the burning of fossil fuels. None of these have survived serious peer review, but one has more than a grain of truth. This is
that much of the exceptional warming, and the extreme weather
experienced from 2014 till 2016 has been caused by the El Niño
effect in the Pacific Ocean off the coast of South America. El
Niño is a natural weather pattern that occurs when ocean temperatures in the Pacific Ocean near the equator start to rise above
the normal. In years when this does not happen, strong trade
winds blow from east to west across the Pacific Ocean around
the equator. These push warm surface ocean water from South
America west towards Asia and Australia, and cause cold water to
well up from the ocean depths in the east to take its place along
the South American coast. This creates a temperature disparity
across the Pacific, which reinforces the trade winds' flow from
east to west. The accumulation of warm water in the west heats
the air above it and loads it with moisture. This is the genesis of
the typhoon season that regularly occurs across the China–Japan
littoral in the late summer.

Every once in a while, however, the temperature anomaly in the
eastern south Pacific becomes strong enough to be designated an El
Niño. The phenomenon typically occurs every two to seven years.[16]
But the El Niño too shows a cyclical pattern. Every 14 or 15 years an
exceptionally large temperature anomaly creates a 'super El Niño'.
The 2015–2016 El Niño has qualified as one such, for it has been
the worst in 15 years. The two previous super El Niños occurred
in 1982–1983 and 1997–1998. Climate sceptics have, therefore,
seized upon this to dismiss the exceptionally hot and violent weather
of the last three years as another natural phenomenon being misused
by the climate science community for their own purposes.

The flaw in this argument, however, is that each of the three
preceding super El Niños has been more violent and has created

greater temperature and weather extremes than its predecessor. The term super El Niño was coined to describe the El Niño of 1982–1983. When an even stronger El Niño came along in 1997–1998, it was termed the El Niño of the century. The El Niño of 2014–2015 has simply been called 'monstrous'. The rise in the peaks of the three super El Niños is in fact additional evidence of the acceleration of global warming and climate change that has set in.

The prime cause of the ever stronger super El Niños too seems to be the accelerating accumulation of CO_2 and GHGs in the atmosphere. An international collaboration of 49 scientists working on the GCP reported that between 1990 and 2013, CO_2 emissions had risen by an astonishing 61 per cent.[17] What was more alarming, the rise had been exponential: emissions rose by 29 per cent between 2000 and 2008, a shade over 3 per cent a year,[18] but rose by a staggering 5.9 per cent in 2010, the largest increase on record.[19]

The only way to retain even vestigial control of the global warming process is to bring human CO_2 emissions down well below the 18 to 20 billion tonnes of CO_2 that the earth's natural sinks can absorb before the middle of this century and altogether eliminate anthropogenic carbon emissions some time before its end. But the entire industrialised world is headed in the opposite direction. The most recent example is the welcome that has been given to 'fracking' as a way to reduce CO_2 emissions by replacing coal as the primary source of energy with natural gas. All of the dangers this involves, such as the pollution of drinking water aquifers and the creation of vast chemical lakes and mountains of sulphur that poison hundreds of square miles of land around them, have been quietly pushed into oblivion. James Hansen's derisive comment on the Hallelujah chorus celebrating this 'breakthrough' is worth repeating: 'these governments seem to have forgotten that when there is already too much CO_2 in the atmosphere, slowing down its further accumulation will at most postpone the day of reckoning, not avert it'.

What is the tipping point?

Climate modelling has taken giant strides since the threat from global warming first surfaced in the 1970s. But it is still not an exact science. This is apparent from the wide margins of possible error that surround every central quantitative estimate that the IPCC's fifth assessment report contains. So it is possible that the six tortured years that the planet has just passed through will be followed by a few years of relative calm as El Niño is replaced by La Niña in the pacific. But what will not slowdown without determined and concerted global action is the accumulation of CO_2 and other GHGs in the atmosphere. If this continues then, sooner or later, the earth will cross a tipping point beyond which global warming will set off a succession of changes in the earth's biosphere that will make global warming self-reinforcing and take it out of human control.

The earliest, and still the most compelling, formulation of the theory that the earth is close to a tipping point was by the celebrated British scientist, James Lovelock. Lovelock's work is seldom referred to in the current debate on global warming because of the mystical overtones it acquired in the 1980s. In 1979, he pointed out that the earth was not just a semi-molten chunk of matter cast off from the sun that happened to go into orbit around it at the right distance for a crust to form and trigger the sequence of events that have led to the creation of sentient life. On the contrary, life has not only been shaped by but has also shaped the evolution of the earth's ecosystem. Today, three billion years after the birth of life on the planet, its fauna and flora, its atmosphere, its forests, its ice and snow cover and its sea and land temperatures are not governed by separate ecosystems but are part of a single interdependent system whose stability depends upon its not being disturbed. Every disturbance of one of its elements causes the rest of the components to adjust in order to restore equilibrium. If the disturbance of the equilibrium is sufficiently powerful and sustained, he argued, at a certain point the earth will adjust all of its parameters to establish a new equilibrium. The change from the old to the new equilibrium would not be gradual, he predicted, but would probably take place within a few decades. In this sense, the earth would behave very much like a living organism that fights back to repel invaders.

The driver of all change in the earth's biosphere, he wrote, is the energy it receives from the sun, and the sun has been getting hotter with the passage of time. Three billion years ago it was a much younger star and gave off a good deal less heat than it does today. The earth therefore received about a quarter less energy from it than it does today and was a rather cold and dark place. Its atmosphere contained 30 times as much carbon dioxide as it does today. There was very little oxygen. The earliest living organisms developed in the absence of oxygen, and many are still around as anaerobic bacteria that only have to be exposed to oxygen to die. In that early earth, the decay of these organisms created methane which is, overall, 23 times as potent a GHG as CO_2. The blanket of methane created a huge greenhouse effect, but on the cold, dark, planet this was a benevolent development that aided the evolution of complex forms of life, for it kept the heat generated by the decay of dead matter, from escaping into the space.

As the sun aged and grew hotter, the need to conserve the earth's own heat declined. Temperatures on the earth's surface rose and the thick mantle of CO_2 and methane became a hindrance. Aerobic—meaning oxygen-breathing—living matter began to multiply. The miracle that solved this problem was photosynthesis. This steadily lowered the concentration of CO_2 and raised that of oxygen. It began with the growth of algae in the oceans and, as the oxygen content of the atmosphere rose, spread rapidly to the land. Lovelock christened this active, almost sentient, earth 'Gaia'.

In Lovelock's estimation, the earth was healthiest about a billion years ago. At that point, the sun's heat was just sufficient to make it unnecessary to either conserve or disperse any of it. But as the sun continued to grow hotter, the earth had to increasingly manufacture the conditions in which the verdant and abundant life of that period could continue to flourish. It did so by generating ice ages. These had several effects: they cooled the earth by increasing the size of the ice caps, and they cooled the oceans and revived its rich plant life, thereby increasing its capacity to sequester carbon. They reduced the land area that could sustain life, but also exposed large amounts of fresh new land for the spread of flora and fauna by lowering the sea level. In the last ice age, it was lower by as much as 120 m.

The proof that the ice ages restored the earth's health is to be found in the very low concentration of CO_2 in the atmosphere during the last one—180 ppm. When it ended, the CO_2 concentration began to rise and stabilised around 280 ppm, where it stayed till around 1850 AD. This was 120 ppm lower than the level that prevails today. As Lovelock notes, it must have needed a lot of vegetation before the last ice age to pump the CO_2 concentration down so low.

But as the aeons have passed and the sun has continued to grow hotter, 'Gaia's' capacity to regulate itself has begun to approach its limit. Lovelock estimates that even without human interference the earth has about a hundred million years to go before it becomes a dead planet. Human intervention has therefore come at a time when its capacity to adjust is close to being exhausted.

Lovelock identified two tipping points: the melting of the polar ice—summer ice in the arctic and the Greenland west Antarctic ice sheets—and a warming of the oceans to the point where the sea algae that populate vast areas of water close to their surface begin to die. The first of these is already upon us. The appearance of more and more open sea in the Arctic, free from ice during the summer, is reducing the area of the earth's surface that reflects the sun's rays back from ice into space and increasing the area in which it is absorbed by the dark waters of the sea. This is heating up the water and speed up the melting of the Arctic ice, widening the area exposed to the sun's rays in summer more and more with every passing year.

The warming of the oceans upsets a delicate balance between two opposed tendencies in the upper layers of the ocean. At its surface, the evaporation of the water increases its salt content and makes it denser. The increased weight pushes it down and brings colder, nutrient-rich water up from the lower depths. But as the surface water warms, it also expands and becomes less dense. Beneath a threshold temperature, evaporation wins and sends the warm water down. The upper layers therefore remain full of fish and other marine life. But once the surface ocean temperature crosses that threshold, the surface water ceases to sink down and does not therefore force the colder water below to rise to the surface. Since this 'overturning' is needed to replenish the supply of nutrients in the uppermost layers of the ocean where the algae and plankton live, the warming of the ocean kills the phytoplankton that fix the CO_2 through photosynthesis. Since the oceans account for around 11 billion tonnes, or two-thirds of all the CO_2 captured by nature, the danger that the warming of the seas represents is even greater than that posed by deforestation, and it is one that human beings are only now beginning to understand.

In 1994, Lovelock and geochemist Lee Kump published an article in *Nature* that attempted to understand the extent of coupling and feedback between plant growth, global temperature and enhanced atmospheric concentrations of GHGs with the help of an elaborate computer model that included just about every variable they could think of. These included the behaviour of ice caps, tropical forests, algae and small periodic variations in the heat of the sun. In the computer simulation, the concentration of CO_2 was increased to three times the pre-industrial level of 280 ppm very gradually over a period of 20,000 years (a rate far slower than the pace of accumulation today).[20] It gave some surprisingly prophetic results. As the concentration of CO_2 approached 500 ppm, it pushed up the mean air temperature by about 3°C. Two things then happened at almost at the same time: the Greenland ice cap became unstable and the changes in land vegetation and ocean algae became sharply more pronounced, suggesting that a self-reinforcing mechanism had been set in motion. At 500 ppm, the melting became irreversible and the capacity of the sea algae to absorb CO_2 nosedived. In their model, once the earth crossed this tipping point, the mean temperature rose within decades by 16°C to 24°C! In Lovelock's assessment, the situation in 1994 was similar to that which had preceded the sudden onset of an extremely hot period fifty-five million years ago, called the Eocene period, when the earth suddenly warmed by 8°C in the temperate regions and 5°C in the tropics.

Geological studies by Professor Harry Elderfield at Cambridge University suggested that the Eocene period was triggered by a sudden release of somewhere between 0.3 and 3 trillion tonnes (terra-tonnes) of fossil carbon into the atmosphere. Human beings had already added half a trillion tonnes of carbon to the earth's atmosphere in the previous hundred years, and the rate of accumulation was rising exponentially. Lovelock therefore warned his readers that the earth's temperature had entered the range that had triggered the onset of the Eocene hot period.[21]

In his book *Gaia*, Lovelock predicted that the adjustment of the earth's parameters after it crossed the tipping point would not be gradual but sudden. His extraordinary prescience can be gauged from the fact that *Gaia* was written in 1979, 'a full decade before the study of ice core samples from the Greenland ice sheet began'. These have since confirmed that in past ages, climate change had indeed been sudden and not gradual. The end of the ice age, 15,000 years ago, was one such event. The descent into a 'little' ice age, known as the Younger Dryas period, 12,800 years ago and the ascent out of it 11,500 years ago were both equally sudden. In the highest areas of the Greenland ice cap, temperatures fell by 15°C in a matter of one or two decades when it began and rose by around 10°C in 40 to 50 years when it ended. Some recent findings suggest that the rise in temperature at its end might have been even faster.[22] There was another 'cold event' 8,200 years ago that showed a similar pattern.

Twenty years after Lovelock and Kemp's simulation, it is disturbing to see how closely reality has tallied with their predictions. In the past century, the average temperature of the atmosphere has risen by about 1.0°C. This may not look like much but the global average is a deceptive measure because it hides a two to three times greater increase in the Arctic region. In this region, stretching from Alaska, across the Canadian to the Siberian tundra, the increase in temperature has been 2°C to 3.5°C already.[23] It is this that has melted more and more of the ice that covers the Arctic in the summer and is beginning to melt the permafrost in Alaska, Siberia and Russia.

An even more disturbing vindication came from a study of CO_2 absorption in the Sea of Japan. In 2007, a joint South Korean–Russian study team took a large number of samples of sea water and compared their CO_2 content with that of samples taken in 1992 and 1999. The scientists found that between 1999 and 2007, the sea had absorbed half as much CO_2 as between 1992 and 1999. The cause, according to Kitack Lee of the Pohang University of Science and Technology, was a weakening of the vertical circulation of the sea water. This had reduced the density of the sea algae in the upper layer of the sea and consequently its capacity to trap CO_2.

This was the first study of its kind and covered only an 8-year period, so many scientists initially treated its findings with caution.[24] But only two years later, a study of CO_2 absorption by the sea from 1765 to 2008, by a team of scientists led by Samar Khatiwala from the Lamont–Doherty Earth Observatory of Columbia University, confirmed that this was indeed happening. It found that starting in 1950 there had been a large increase in the percentage of CO_2 emissions that the seas were absorbing, but after 2000 this started declining, although in absolute terms the tonnage absorbed continued to increase.[25] Two years later, another study, this time by scientists at the University of Wisconsin–Madison confirmed Lee's finding that over large swathes of its surface the ocean's ability to absorb CO_2 was also falling in absolute terms.[26] The less the sea absorbs, the more rapid will be the concentration of CO_2 in the atmosphere increase. These studies are a grim warning that Lovelock's tipping point in the oceans may already have been crossed.

James Hansen's warning

In 2008, a powerful warning of imminent, catastrophic climate change was issued by James Hansen, whose testimony to the US Congress 20 years earlier had triggered concern in the United States over the perils of global warming. In sharp contrast to the carefully hedged warning issued by the IPCC in its fourth assessment report (2007), Hansen did not hesitate to spell out the conclusions that were implicit in its findings. Chief of these is that the goal accepted by the world's leaders for 2050, of keeping the rise in air temperature down to an 'average' of 2°C above pre-industrial times will mean a 4°C to 5°C rise in the arctic region.[27] This would be sufficient to trigger self-reinforcing global warming that would be out of human control. The battle to keep the world as it has been for the past 7,000 years and as we know it today will have been lost.

In language that has a touch of poetry, he told the US Congress in 2008:

> The concentration of CO_2 in our atmosphere now stands at 387 parts per million, the highest level in 600,000 years and more than 100 ppm higher than the amount at the dawn of the Industrial Revolution. Burning just the oil and gas sitting in known fields will drive atmospheric CO_2 well over 400 ppm and ignite a devil's cauldron of melted icecaps, bubbling permafrost, and combustible forests from which there will be no turning back.... More warming is already in the pipeline, delayed only by the great inertia of the world's oceans. And the climate is nearing dangerous tipping points. Elements of a perfect storm, a global cataclysm, are now assembled.[28]

Hansen has been called an alarmist more times than he can remember. But he is and always has been a scientist. He made his name, while still a young researcher, when he and his colleagues showed that Venus owed its brilliance not to a cloud of water vapour around it but one of sulphuric acid. In the 1990s, while he was issuing increasingly anguished warnings of the threat from global warming, he was looking diligently for doable ways in which it could be contained within a relatively short span of time. Aware that since viable alternatives to fossil fuels did not yet exist and reducing CO_2 directly was not going to be easy, he and his colleagues suggested that the US government concentrate its efforts on reducing the emissions of other GHGs, notably methane, which together were responsible for half of the human-induced global warming.[29] But during the Bush administration, his pleas fell on deaf ears. Still later, in the beginning of the last decade, he published a study of how aerosols—particulate matter and gases thrown up into the atmosphere—had acted as a shield against the sun's rays in the past and could do so in the future if the government controlled the so-called 'black aerosols', which absorbed and trapped the suns heat and encouraged the increase of 'white' aerosols which did the opposite, in the upper atmosphere.

He decided to shed his scientific reticence only in the second half of the last decade when the evidence gathered by tens of thousands of scientists, with instruments that had not even been deployed a decade earlier, began to come together to form a picture that was sufficiently alarming to deny him the refuge of doubt.

The logic behind his decision was simple and therefore irrefutable. An array of new measuring devices placed in the upper atmosphere and on the surface of the sea had made it possible to measure the amount of radiation being received and sent out by the earth every day with remarkable accuracy. These showed, beyond any possibility of doubt, that the earth was absorbing more energy from the sun during daylight hours than it could radiate out at night. Something had to give. Sooner or later, he concluded, the earth would re-establish the equilibrium. To do so it would have to radiate more heat. And to do that it would first have to become hotter.

Hansen therefore came to exactly the same conclusion that Lovelock had come to, except that he did so on the basis of measurements by instruments that had not even been thought of when Lovelock developed his concept of a sentient Earth. Nine-tenths of the excess energy, Hansen warned, is being absorbed by the sea. Because the thermal capacity of water is huge, and the sea covers three quarters of the earth's surface to an average depth two-and-a-half miles, this has only now begun to make a discernible impact on its temperature. Recorded data show that sea-level rise has accelerated from 1.23 mm a year between 1953 and 1993 to 3.1 mm a year between 1993 and 2003.[30] About half of this has been contributed by the expansion of the sea water and a slightly smaller amount by the faster melting of various glaciers, and of the Arctic and Antarctic ice.

Sea-level rise—IPCC's flawed model

When, based on these figures, the IPCC forecast a rise of between 18 and 59 cm by 2090–2099, most policy makers heaved a sigh of secret relief. True, policy makers thought but did not say, a 60 cm rise will drown many pacific atolls and cause some tiny Island states, like the Maldives, to virtually disappear. But their population is small and can be accommodated on the nearest mainland.

Hansen pointed out, however, that this projection was severely flawed, but had been accepted by the IPCC in 2007, in spite of that. The most serious was that it had only taken the rise in sea level caused by the expansion of water caused by rising temperature into account, and left the addition of fresh water caused by the melting of the ice fields.[31]

Scientists had noticed that while the sea had been warming at a consistent pace between 1953 and 2003, the rise had virtually stopped after 2001.[32] In an article they published in *Science* in April 2010, two scientists Kevin E. Trenberth and John T. Fasullo concluded that the reason had to be that a large part of the energy being received by the sea from the sun is going into converting the Arctic and Antarctic ice into water. Changing a single gram of ice into water *at the same temperature* requires an input of 80 calories (334 joules) of energy. This is sufficient to raise the temperature of a single gram of water by 80°.[33]

The IPCC did not take this effect seriously into account because it was following a flawed model of how the ice in the polar and Arctic ice sheets actually melted. The prevailing belief in the climate modelling community when it prepared its fourth assessment report was that while global warming could easily increase the melting of the polar ice sheets at the edges, the increase in water vapour in the air, caused by the rise in temperature, would lead to heavier snowfalls over the entire ice sheet in the winter. This would largely offset the fraying at the edges. Thus, some models had even concluded that global warming would increase the size of the polar ice sheets.

Hansen pointed out that it failed to account for the actual behaviour of the two ice caps over geological time. Paleo-climatic history, gleaned from the Greenland ice core samples, showed that on several occasions rising temperatures had reduced the size of the ice sheets so fast that it had raised the level of the sea by several metres in a century. During the last ice age, the sea level was fully 120 m below its present level. Sea level changes of up to a metre in a decade had been fairly common. In fact, Hansen pointed out, it was the stability of the sea level during the past 7,000 years that has been something of an anomaly, for it has required a delicate equilibrium in which the world was warm enough to prevent the Canadian and Siberian ice sheets from expanding southwards, yet cool enough not to melt and shrink.

There had, therefore, to be some other reason for the large and rapid changes that had occurred in earlier times. Hansen concluded that the energy needed to melt so much ice in such a short time had to be coming from the sea. He calculated that if the sea were to transfer all of the surplus energy that it will receive in the rest of this century into the melting of ice sheets, and its temperature remained unchanged, the sea level would rise by *5 to 6 metres* by 2100, and the rise would continue into the centuries that follow. But if the concentration of CO_2 and other GHGs in the air continues to rise and the current energy imbalance, of $0.6\,W/m^2$ of the earth's surface, rises further, the sea will, inevitably, transfer more and more of the energy it receives from the sun into the melting of the ice caps and ice sheets at the poles. And it will do so in a shorter and shorter period of time. This is the self-reinforcing feedback cycle that climate scientists dread and the world continues to ignore.

Ice melt accelerates

There is now no room left for doubt that the melting of the polar ice is accelerating. On 12 December 2012, a research team from 'Extreme Ice', an organisation set up by photojournalist James Balog, filmed $7.4\,km^2$ of ice calve off the Ilulissat glacier in Greenland and fall into the sea. It was the largest ice-calving event ever recorded and became a part of the film *Chasing Ice* which was released a short while later. 'It's like watching Manhattan break apart in front of your eyes', one of Balog's researchers said.[34] What *Chasing Ice* did was to destroy deniability by bringing this cataclysmic event into our living rooms.

The acceleration of ice melt is happening before our eyes, and the ice is not melting solely at the edges of the ice sheets but all over them. This is because ice sheets are not flat but have peaks and depressions, and are ridged, often in starkly beautiful ways. Thus, when the summer sun heats their surface, the water does not simply flow off the edges into the sea. Instead it runs down the valleys between the ridges to the lowest points between them and, once there, begins to drill a hole straight down through the ice to the sea below. These holes are called *moulins*.[35] The greater the rate of melting, the faster the flow of water through the *moulin*, the more heat does it transfer to the surrounding ice on its way to

the bottom. At the bottom this cold water does not mix easily with the warmer water under the ice sheet—a peculiar characteristic of water, but slides down along the bottom of the ice sheet till it reaches the sea. It thus acts as a lubricant that speeds up the descent of glaciers towards the sea. On its way along the bottom, the water also encounters ice shelves that reach down from the glaciers into the solid earth underneath and act as anchors that slow down their descent towards the sea. By eating through them the water breaks their moorings and further hastens their downward slide. The two effects together are leading not only to a faster descent of the Greenland and Newfoundland glaciers towards the sea but also to the much higher rate of calving of icebergs and the breaking off of huge chunks that have been caught spectacularly on film in recent decades.

The appearance of clear water in the Arctic early in the year, a dozen years ago, was too obvious and too big an event for the public not to notice. But until very recently the evidence that the rate of melting had risen dramatically and that, consequently, glaciers and ice sheets were losing mass and moving, and breaking up more rapidly, was being obtained only through complex aerial and satellite-based calculations that left the climate sceptics with room to accuse the scientists such as Mann and Hasna in of dishonesty (see Chapter 9). But today, there is photographic evidence that puts the issue beyond doubt.

Four years ago, James Balog, whose film *Chasing Ice* was the official selection for the Sundance film Festival in 2012, started an Extreme Ice Survey in which he set up 30 still cameras in protected locations overlooking some of the largest glaciers in Greenland and Alaska, and programmed them to take photographs every few minutes through night and day, summer and winter for more than three years. He then ran them together to create a unique time-lapse film that shows their movement through the 3 years. The resulting film leaves no room to doubt what is happening to the ice, for by speeding up time it shows glaciers growing in winter and shrinking in summer, but each year doing a little less of the first and a little more of the second; moving faster every summer and, finally calving into huge icebergs at an accelerating pace. Balog's photo odyssey actually shows the Columbia glacier in Alaska retreating 1.95 miles in 3 years and the Llulisat glacier in Greenland, the largest glacier in the world by a huge margin, doubling its rate of descent into the sea from 60 to 125 feet a day.[36]

The speed at which Greenland's ice is melting has also been captured by a video movie, *Melting Moulin,* made by The Telegraph, London, in 2009. It shows two scientists, Jason Box and Basil Springer, taking a reading of the speed at which water is gushing into a *moulin* in the Greenland ice sheet. They recorded a speed of 9.4 miles (15 km) an hour and an outflow from this one *moulin* of 42 million litres a day. At this rate, Greenland is losing enough water every summer to cover Germany with a metre of water.[37] More recent studies confirm that this is indeed happening. The rate at which several other glaciers in Greenland are moving down to the sea has also doubled and in some cases tripled in the past decade.[38] Nor is the acceleration confined to Greenland: the area of the Arctic covered by permanent ice that is more than five years old, fell from 57 per cent in 1987 to a paltry 7 per cent in 2007.[39] Scientists measuring the water temperature under the tip of the Sermilik Glacier in November 2010 found it to be as high as +5°C (40°F). The ice, they realised, was continuing to melt from the bottom even after it had stopped melting, at the end of summer, from the top.

Scientists working on Project Clamer, a collaboration of 17 institutes in 10 European countries that has been synthesising research from nearly 300 European Union (EU)-funded projects over the past 15 years, have warned that the accelerated melting of the sea ice and the surrounding glaciers is creating a huge 'lake' of fresh water, that is water too light to 'overturn' in the Arctic. In April 2011, even before the summer had begun, it was twice the size of Lake Victoria and growing. Laura de Steur, an oceanographer at the Royal Netherlands Institute for Sea Research, warned:

The volume of water discharged into the Arctic Ocean, largely from Canadian and Siberian rivers, is higher than usual due to warmer temperatures in the north causing ice to melt. Sea ice is also melting quickly. The permanent ice cover of the Arctic ocean reached a record low in September 2007 when it covered barely 4.1 million km^2 of the Arctic's 14 million km^2. This was followed by a new record low in 2012 and further declines in 2014 and 2015. This has added even more freshwater to the Arctic Ocean.

The shrinkage of the ice cover is, however, only one consequence of global warming. A second one whose importance has been appreciated only recently is that the remaining ice cover of the Arctic is now much thinner than before, and the fresh ice created over vast tracts in winter is even more so. It is therefore being broken up by winds that are gaining in strength because of the opening up of the sea ice and consequent warming of the ocean. These winds have broken the earlier pattern of rotary circulation of the floating ice within the Arctic, and are driving it out into the Atlantic. That is lowering the salt content of the extremely frigid water in which the ice sheets form, and slowing down the sinking of the cold, salt-heavy waters in the North Sea that initiates the thermohaline current. What will happen when this gains momentum is hard to predict.[40]

The meltwater is particularly destructive when ice sheets are anchored to the seabed below by ice shelves. In the West, Antarctic Ice Sheet in particular, the meltwater is dissolving these shelves and 'freeing' the ice, which has begun to break away in country-sized chunks. In March 2008, a $400\,km^2$ piece of the West Antarctic ice shelf (the size of Connecticut) broke away. In February 2009, a giant, $14,000\,km^2$ chunk of the Wilkins shelf broke away and floated northwards, leaving a bare $2,000\,km^2$ still attached to Antarctica. In November 2011, a $750\,km^2$ chunk of the Pine Island shelf broke away.[41]

A European satellite Envisat, which has been watching the North Antarctic peninsula, has reported the loss of $5,000\,km^2$ from the Larsen Ice Shelf between 2002 and 2012. The Larsen Ice

Shelf consists of three shelves that line the eastern side of the North Antarctic peninsula. Only 15 per cent of the shelf that existed in 1995 now remains. The peninsula, which juts out north into relatively warmer zones, is one of the fastest warming areas in the world, for it has experienced a 2.5°C rise in temperature in the past 50 years, that is, more than three times the global average rise. In July 2017, a colossal ice-berg, measuring about 2,200 square miles, the size of Delaware state in the USA, splintered off the 'C' section of the shelf.[42]

The decline of the Albedo effect

But the mortal threat to humanity and most other living species on earth comes not from the Antarctic but the Arctic. This is because of the rapid and accelerating decline of a crucially important regulator of the earth's temperature called the Albedo effect. The Albedo is the proportion of radiated energy received that is reflected back into space. It is highest in the polar regions in winter because the ice and fresh snow cover reflect 90 per cent of the sun's heat back into space. As the polar regions warm, ice sheets break up, lowering its Albedo. As more of the surface of the earth gets exposed to the sun it begins to warm. This process is now so far advanced that it has begun to melt the permafrost. The permafrost is a layer of the earth's crust, varying in depth from several metres to two miles that stretches in a wide belt around the North Pole and underlies the Arctic Ocean. Within it is trapped an estimated 1.3 to 1.7 trillion tonnes of carbon in the form of decaying vegetation that was frozen at the onset of the last ice age, 30,000 years ago. This is more than 2,000 times the annual, human-induced, emission of carbon, in the form of GHGs. The melting of the permafrost could become the time bomb that pushes the world over the edge of disaster.

The permafrost is already thawing in summer. This is visible across the entire northern sub-Arctic region, from Alaska to Siberia. Although it began barely a decade ago, it is already having bizarre effects. Roads and railways built on what seemed to be solid ground have developed cracks; lakes in Alaska have suddenly run dry as the water has seeped away; gas is bubbling up from swamps in the Russian tundra and holes cut in the ice on frozen lakes have shot out jets of flame. The evidence is thus mounting that the permafrost is beginning to release some of the carbon that is trapped within it. So far most of the release of GHGs has taken place on the land and in shallow marshes. As a result, most of it is CO_2, although a significant part (about 2.7 per cent) is methane. This makes the GHGs released so far about 2.5 times as potent an agent of global warming as the CO_2 released by the burning of coal or wood.

The amount of methane being released is at present small, but will rise catastrophically if the Arctic continues to warm at the present rate. Unfortunately, global warming is all set to make it do precisely that in the coming decades. Even if the increase in mean air temperature is contained at 2°C by 2100, which is now next to impossible without an early and complete shift out of fossil fuels, the two to threefold greater rise of the mean air temperature in the Arctic region will greatly accelerate the release of gases from the permafrost. And a far higher proportion will be methane. While estimates are still tentative, there is a consensus among 41 scientists working as part of a worldwide Permafrost Carbon Network that at the present rate of consumption of fossil fuels the air temperature in the arctic will rise by 2.5° higher by 2040 and 7.5° by 2100. Should that happen, the permafrost is likely to release between 30 and 60 billion tonnes of carbon as methane and CO_2 by 2040 and 232 billion to 380 billion tonnes by 2100.

Even if the world gets its act together in time to prevent a further increase in temperature after 2100, the permafrost will continue to release carbon into the atmosphere for centuries in the future. For just as the thaw has taken a century of global warming to initiate, it will take several centuries for a sustained reduction of CO_2 emissions to reverse the process. The PCN estimates that the release will reach 549 billion to 865 billion tonnes by 2300. The thawing of the permafrost can therefore increase man-made emissions of GHGs by a third in the rest of this century and double the rate of concentration of CO_2 in the atmosphere.[43]

Clathrates—The ultimate threat

Serious as the threat from the melting of the permafrost is, it is dwarfed by the melting of Clathrates.[44] The Arctic covers approximately 14.6 million km^2, and all of it rests on a bed of frozen decayed matter. Of this till 1970, around 8 million km^2, or almost half, used to be covered by permanent ice throughout the year. In the rest of the ocean, the ice would break up in summer, forming floating ice packs that would move around in the ocean till the onset of winter made it freeze again. Global warming has disturbed this balance. From 1970 or thereabouts, the area under permanent ice has begun to decrease. In September 2005, it hit a till then all-time low of 5.3 million km^2. Two years later it fell to 4.1 km^2.[45] In September 2011, it again hit a near-record low of between 4.21 and 4.33 million km^2.[46] Taken together, the last 12 years have seen the lowest permanent ice cover in the past half a century. The warming of the Arctic is proceeding far more rapidly than scientists had originally expected. The models constructed in the early 2000s and used by the IPCC for its fourth report had predicted that the Arctic Ocean would be free of summer ice by 2070–2100, but evidence has mounted steadily that shows it could happen as early as 2020, just three years from now.[47]

    With the heat absorbing capacity of ice-melt no longer available during the summer months, the temperature of the surface layers of water will start to rise rapidly. The average depth of the Arctic is only just over a thousand metres, but the continental shelves that ring it are much shallower, attaining average depths of just 40 m. It will not take long, therefore, for the heat to reach the permafrost underneath, and its deadly load clathrates.

The areas where the summer ice has retreated are, by defini-
tion, the continental shelves in the Arctic where the ocean floor is
closest to the sun. This vast area will therefore be the first to start
releasing methane, CO_2 and other gases into the atmosphere. In
A Farewell to Ice, Peter Wadhams has estimated that if only 10 per
cent of the methane trapped in the clathrates of the East Siberian
continental shelf were to be released, it would increase the global
mean temperature by 0.6°C more than current predictions, as early
as in 2040. This is the fourth tipping point, in addition to the
three identified by Lovelock and Hansen, that could decide the
fate of humanity, and it can do so very soon.

The amount of carbon bound in gas hydrates is conservatively
estimated to total twice the amount of carbon to be found in all
known fossil fuels on earth. The study of the Arctic seabed and
the permafrost-bound land around it is still in too early a stage to
be able to predict how soon this will happen. But if the warming
of the Arctic ocean region continues unchecked, it shall happen.
Indeed, given the fact that the air temperature in the Arctic is rising
at least twice as fast as in the world as a whole, and that a 3° rise in
the latter over pre-industrial times by 2050 is almost impossible to
prevent, we may have crossed this tipping point already.

Tipping point—What could happen by 2100?

If global warming continues, even at a reduced pace, all these
feedback effects and others that have not been described here for
reasons of brevity will come together by the end of this century,
if not sooner. The Arctic will be clear of ice in summer, and the
entire 14,600 km² will be absorbing nine-tenths of the heat from
the sun instead of reflecting two-thirds back into space. The release
of GHGs from the permafrost and the Arctic bed maybe adding
another 1.5 to 3 billion tonnes of carbon (the C part of the CO_2)
to the 9 to 10 billion tonnes we are adding today. So far only
2.7 per cent of the gas released by the ice melting of the permafrost

has been methane, but this ratio will change dramatically when the clathrates start melting, for the gas rapped inside them is pure methane. Peter Wadhams has estimated that if only 10 per cent of the clathrates in the East Siberian continental shelf melt, the resulting methane pulse will add 0.6°C to the rise that is already in the works for the period 2020 to 2030.[48]

In the months before the Copenhagen Climate Conference of 2009, a consensus had developed in the Climate Change Community that, to keep the rise in temperature during the 21st century below 2°C the concentration of CO_2 had to be kept below 450 ppm in 2100 AD. This was formally adopted as an aspirational goal at Copenhagen and reaffirmed at the Paris summit in 2015. But that target was already well beyond reach in 2009 and is completely unattainable today. The concentration of CO_2 reached 400 ppm in 2012 and has been rising by a fraction over 2 ppm a year. If this continues through the 21st century, by its end, the concentration of CO_2 will exceed 550 ppm, and the average rise in temperature by 2100 AD will be not 2°C but 3 to 3.5°C. Should that happen, the world will, beyond a shadow of doubt, have crossed all the tipping points Lovelock had identified, and global warming will not only be accelerating but completely out of human control.

A 3° warmer planet

What will a 3° warmer planet look like? There is, as yet, no global simulation model that describes what will happen to every region of the world. But detailed modelling has been done for parts of it. In his 2007 book *Six Degrees: Our Future on a Hotter Planet,* Mark Lynas has used these to describe what is likely to happen to several parts of the planet in vivid detail. Three degrees of warming by 2100 is likely to split the African continent in two, turning the lower half into a hyper-arid desert once again, while 'greening' some of the upper parts with increased rainfall. In the southern part, rainfall will decline by 10 to 20 per cent, prolonging droughts and making them more frequent. The droughts will combine with a doubling of wind speeds caused by the warming of the oceans around southern Africa, to untether the giant sand dunes which are presently stabilised by savannah grasslands and scrub vegetation. By 2040, the dunes of the southern Kalahari, which cover Namibia and Botswana, will start moving. By 2070, the dunes of the northern Kalahari, in Angola, Zambia and Zimbabwe are also likely to get untethered. By then, all computer simulations agree, almost all of Botswana is likely to be covered by moving sand dunes. Over the next several hundred years the cumulative effect of 3° of warming will be to desertify the whole of the American south-west and all of Pakistan, desiccate most of northern India, make the poles free of ice and spread vegetation to till within a few score kilometres of both.

The evidence is to be found in paleo-climatic studies of what happened during the Pliocene period, three million years ago, when the main contours of the earth were pretty much the same as they are today, and the mean air temperature was about 3°C warmer than it is today. Two studies, in 1995 and 1997, showed the presence of fossil wood from stunted shrubs in rock sediments only $500 km^2$ from the South Pole, and similar fossilised wood from pines and other coniferous trees several hundred kilometres closer to the North Pole than any trees grow today. A third study in the far north of Canada unearthed animal remains that enabled researchers to determine that the temperature had been 15°C above what it is today and that birch and larch had grown 2,000 km north of the present tree line. Subsequent studies showed, beyond doubt, that the cause was not a warming of the oceans, as some scientists had surmised earlier but a high level of CO_2 in the atmosphere.[49]

What is most disturbing about his extrapolation is that the concentration of CO_2 that had resulted in this degree of warming was only 360 to 400 ppm. We have already crossed this limit, for in November 2016 the Mauna Loa Observatory in Hawaii recorded it as 403.64 ppm.[50] So we have already set the stage for the changes that occurred during the Pliocene period. What a 550 ppm CO_2 concentration by 2100 AD, which is where we are headed today, will do to the planet when its full effect has been absorbed is too hideous even to contemplate. Humanity is therefore living on borrowed time.

Notes and references

[1]　*Financial Times*, 'North Pole's November Heat Wave raises Climate Fears' (London, 23 November 2016), 1.

[2]　BBC World News Saturday, 24 December 2016.

[3]　See https://www.ncdc.noaa.gov/cag/time-series/global/globe/land_ocean/ytd/5/1980-2016 (accessed on 7 July 2017).

[4]　Kelly Levin, 'Timeline: Extreme Weather Events in 2012', World Resources Institute (6 September 2012).

[5]　See http://www.thehindu.com/sci-tech/energy-and-environment/chennai-floods-due-to-climate-change/article7980332.ece (accessed on 7 July 2017).

[6]　These were Rajasthan, Madhya Pradesh, Gujarat, Maharashtra, south Uttar Pradesh, south Haryana, Chandigarh and interior Odisha. Read more at http://www.oneindia.com/india/india-burning-after-bhira-s-world-record-temperature-met-2389002.html (accessed on 7 July 2017).

[7]　See http://www.oneindia.com/india/india-burning-after-bhira-s-world-record-temperature-met-2389002.html (accessed on 7 July 2017), and Jacob Koshy, 'Brace for Heat Waves, States Told', *The Hindu*, available at http://www.thehindu.com/sci-tech/energy-and-environment/states-get-warning-to-brace-for-heat-waves/article17710875.ece (accessed on 7 July 2017).

[8]　James Hansen presented this estimate at a speech at the University of Iowa in 2011.

[9]　The northern hemisphere hockey stick graph of Mann, Bradley and Hughes 1999. There are many reproductions. See https://phys.org/news/2015-02-iconic-graph-center-climate-debate.html. Read more at: https://phys.org/news/2015-02-iconic-graph-center-climate-debate.html#jCp

10 Reuters, *Global Carbon emissions rise to new record in 2013*, available at http://www.reuters.com/article/2013/11/19/us-global-carbon-emissions-idUSBRE9AI00A20131119 (accessed on 7 July 2017).

11 See https://www.theguardian.com/environment/2013/sep/27/ipcc-world-dangerous-climate-change and http://climateparis.org/COP21 (accessed on 7 July 2017).

12 James Hansen, *Storms of My Grandchildren: The Truth About the Coming Climate Catastrophe and our Last Chance to Save the World* (New York, NY; Berlin; London; Sydney: Bloomsbury, 2009).

13 See https://public.wmo.int/en/media/press-release/provisional-wmo-statement-status-of-global-climate-2016 (accessed on 7 July 2017).

14 Jason Samenow and Justin Grieser, 'While the North Pole warms beyond the Melting Point, it is freakishly cold in Siberia.' *The Washington Post*, 18 November 2016. Available at: https://www.washingtonpost.com/news/capital-weather-gang/wp/2016/11/18/while-the-north-pole-warms-beyond-the-melting-point-its-freakishly-cold-in-siberia/?utm_term=.b43273adea34

15 See http://newsweekpakistan.com/climate-change-making-extreme-weather-events-worse-u-n/ (accessed on 7 July 2017).

16 For more information, please see State of the Planet, Earth Institute, Columbia University's article 'El Niño and Global Warming—What's the Connection?' available at http://blogs.ei.columbia.edu/2016/02/02/el-nino-and-global-warming-whats-the-connection/ (accessed on 7 July 2017).

17 Reuters, *Global Carbon emissions rise to new record in 2013*.

18 Global Carbon Project, *State of the Global Carbon Budget*, available at http://www.globalcarbonproject.org/global/pdf/GCP_10years_med_res.pdf (accessed on 7 July 2017).

19 By 2013, however, the rise had slowed to 2.1 per cent because of the deepening recession in Europe, a very sharp industrial slowdown in China and the rapid replacement of coal with gas-fired boilers in the United States.

20 J.E. Lovelock and L.R. Kump, 'Failure of Climate Regulation in a Physiological Model', *Nature* 369, no. 6438 (1994): 732–734.
A summary of their article reads as follows:

There has been much debate about how the Earth responds to changes in climate—specifically, how feedbacks involving the biota change with temperature. There is in particular an urgent need to understand the extent of coupling and feedback between plant growth, global temperature and enhanced atmospheric concentrations of greenhouse gases. Here we present a simple, but we hope qualitatively realistic, analysis of the effects of temperature change on the feedbacks induced by changes in surface distribution of marine algae and land plants. We assume that algae affect climate primarily through their emission of dimethyl sulphide (1–8) (which may influence cloud albedo), and that land plants do so by fixation of atmospheric CO_2 (refs 9–12). When we consider how the planetary area occupied by these two ecosystems varies with temperature, we find that a simple model based on these ideas exhibits three feedback regimes. In glacial conditions, both marine and terrestrial ecosystems provide a negative feedback. As the temperature rises to present-day values, algae lose their strong climate influence, but terrestrial ecosystems continue to regulate the climate. But if global mean temperatures rise above about 20 degrees C, both terrestrial and marine ecosystems are in positive feedback, amplifying any further increase of temperature. As the latter conditions have existed in the past, we propose that other climate-regulating mechanisms must operate in this warm regime.

[21]   James Lovelock, *The Revenge of Gaia: Earth's Climate Crisis and the Fate of Humanity* (New York, NY: Basic Books, 2006).

[22]   Takuro Kobayashi et al. '4+/–1.5 Degrees Abrupt Warming 11,270 Years Ago Identified from Trapped Air in Greenland Ice', *Earth and Planetary Science Letter* 268: 397–407.

[23]   IPCC fourth report: Synthesis report.

[24]   David Adam, 'Sea Absorbing Less CO_2, Scientists Discover', *The Guardian* (12 January 2009). Also see http://www.nature.com/articles/nature08526.epdf?referrer_access_token=rAZ4dmJ5R_Z9j8kXN9KHa9RgN0jAjWel9jnR3ZoTv0PlCbjjXOggg2DWo-333wJ8AF5VZIy9exhke_NZ5le_KJ3oRQFIDPqGT3D-eD81NGBFaU70ygFaayTAYOIe8H4UthV0W9Yi3saL3ibnJdCDQ6cqR3ntD7_FaTm06_KaTFelShjfWXBfSA4yK_ncJBKlMBvP1l_vaobm-LBwiZMc98ZYzhCwDehh0RsRx-CIMPYSsCvKunvS9I3Ndv8sN-wu8EQ1Mm1qL45exJgfbIZ6JPy1gwYxeY2Ackh2tuk2lxFZOETY-JkwXlT-rzYca2p_DZH&tracking_referrer=www.treehugger.com (accessed on 7 July 2017).

[25]   *Nature* 462, no. 19 (2009, November).

[26] See http://news.wisc.edu/climate-change-reducing-oceans-carbon-dioxide-uptake/ (accessed on 7 July 2017).

[27] The IPCC's fourth report had pointed out that while average rise should have been 0.8°C during the 20th century, the rise in the Arctic regions had been between 1.5°C and 2.5°C. A 2°C rise in the average air temperature will therefore mean a 4.5°C to 6°C rise in the Arctic region.

[28] Testimony before the US Congress 23 June 2008. James Hansen, presentation on 'Global Warming 20 Years Later', at a Worldwatch Institute event, held on the same day in Washington, DC, 23 June 2008. See Chris Flavin, 'Low Carbon Energy: A Roadmap' (Worldwatch Institute Report No. 178).

[29] For Hansen's own lucid account of his journey from space science into climate science see his book *The Storms of My Grandchildren*.

[30] Calculated by the author. The IPCC figures were 1.6 mm from 1953 to 2003 and 3.1 mm from 1993 to 2003.

[31] For a detailed account of how crucially important the ice-melt has turned out to be and precisely why the ice it has accelerated so dramatically in the past two decades, see Peter Wadhams, *A Farewell to Ice: Report from the Arctic*. Penguin Random House, 2016, Chapter 8, 103–120.

[32] NOAA (National Oceanic and Atmospheric Administration), National Climatic Data Center, State of the Climate: Global Climate Report for March 2012, Tables January to March Global Surface Mean Temperature Anomalies 1880–2012. See chart on April Global land and Sea Temperature Anomalies, available at https://www.ncdc.noaa.gov/sotc/global/201203 (accessed on 7 July 2017).

[33] Kevin E. Trenberth and John T. Fasullo, 'Tracking Earth's Energy', *Science* (16 April 2010): 316–317.

[34] See http://www.theguardian.com/environment/video/2012/dec/12/chasing-ice-iceberg-greenland-video (accessed on 7 July 2017).

[35] French for a wind or water mill.

[36] See http://www.ted.com/talks/lang/en/james_balog_time_lapse_proof_of_extreme_ice_loss.html (accessed on 7 July 2017) and http://www.jamesbalog.com/

[37] Jessica Salter, 'Scientists Capture Dramatic Footage of Arctic Glaciers Melting in Hours', *The Telegraph* (20 February 2009), video footage

available   at   http://www.youtube.com/watch?v=fWInyaMWBY8 (accessed on 7 July 2017).

[38]   Justin Gillis, 'As Glaciers Melt, Science Seeks Data on Rising Seas', *The New York Times* (13 November 2010).

[39]   Alexander Giese, 'Ice Melting Faster Everywhere' (Earth Policy Institute, 24 December 2009), available at www.earth-policy.org

[40]   Fiona Harvey, 'Arctic Freshwater "Lake" to Impact Climate'. Reproduced in *The Hindu* (Chennai, 6 April 2011). Originally published in *The Guardian*.

What actually happens may depend on how fast the 'lake' expands and how rapidly it flows into the Atlantic. If it is large, and sudden, enough it could slow down the thermohaline current and weaken the Gulfstream. Paleo-climatic history records two sudden descents into an extreme cold spell, the Younger Dryas period, from 12,800 to 11,500 years ago and a shorter 'cold event' 8,000 years ago.

There is also a near-consensus that the Younger Dryas 'small ice age' was caused by an overflow into the Atlantic of the huge freshwater lake, dubbed Lake Agassiz, which had formed in the centre of North America when the glaciers of the last ice age retreated. Lake Agassiz contained more water than all the lakes of the world today. Its sudden overflow into the North Atlantic slowed down the thermohaline current and triggered the onset of 'Little Ice Age'.

According to ice core samples taken from the Greenland glacier, both the onset and end of the 'Little Ice Age' took at best half-century. However, Hansen utterly discounts the possibility of another ice age or sudden cooling of the planet. He maintains that human-induced heating is now so great that it will easily overwhelm any event in nature that could have the opposite effect. All that we will be left with will be a sudden rise in sea level measured in metres and not centimetres.

[41]   *Christian Science Monitor*, 4 November 2011.

[42]   Hannah Long, 'Our Antarctica Maps Show the Larsen Ice Shelf's Stunning, Decades-Long Decline'. National Geographic, 17 July 2017. http://news.nationalgeographic.com/2017/07/larsen-ice-shelf-collapse-decline-maps/

[43]   Edward A.G. Schuur and Benjamin Abbott and the Permafrost Carbon Network, 'Climate Change: High Risk of Permafrost Thaw', *Nature* 480, no. (1 December 2011).

44 Also called *methane hydrate, and methane ice*, clathrates are a solid compound in which a large amount of methane is trapped within a crystalline structure of water, forming a solid similar to ice. Originally thought to occur only in the outer regions of the Solar System where temperatures are low and water ice is common, significant deposits of methane clathrate have been found under sediments on the ocean floors of the Earth.

45 Wadhams: op. cit. p. 70.

46 American and Japanese estimates, based on the same observations, differ slightly in their conclusions.

47 Wadhams: op. cit. Chapter 7, 80–84.

48 Wadhams: op. cit. Chapter 9, 126–127.

49 Mark Lynas, 'Six Degrees of Warming' in *Six Degrees: Our Future on a Hotter Planet* (HarperCollins, 2007), Chapter 3, 103–160.

50 See https://www.co2.earth/monthly-co2 (accessed on 12 July 2017).

Chapter Two

A Desperate Search for Remedies

Climate change is not the only threat that we face. More immediate is the one outlined by the Club of Rome in 1972, of systemic economic collapse, followed by famine, pestilence and war. Studies show that 30 years later we are dead on the track of its predictions. So desperate solutions are being sought in various forms of carbon capture. All assume that our dependence on fossil fuels will continue.

Climate change is not the only threat looming over human civilisation. Another is the threat of endless conflict brought on by the steady depletion of non-renewable resources. And this threat has already arrived. The warning had been issued in 1972 by 30 eminent persons who had met in Rome four years earlier to form the Club of Rome. In a report titled 'The Limits to Growth: A Report for the Club of Rome's Project on the Predicament of Mankind', a team of researchers from Massachusetts Institute of Technology (MIT) warned that if human beings continued to multiply, produce, use up mineral resources and pollute the atmosphere at the exponential rate of the previous quarter century, the global economic system would collapse by the middle of the 21st century. Pestilence, famine and war would follow. The most that humanity could hope for would be the establishment of a new, impoverished, equilibrium somewhere near the end of the century.

The MIT research team used a recently developed modelling tool called System Dynamics, which relied heavily on the then new art of computer simulation, to examine the interaction between the growth of population, food production, industrial production, the consumption of non-renewable natural resources and air and water pollution. It concluded that if the trends witnessed between 1900 and 1970 were to continue, the world would overshoot the limits of sustainability somewhere around 2050 AD. The 'correction' that followed would involve a steep decline in population, production and consumption that would flatten out only at the end of the century.

The Club of Rome's warning was the first against infinite growth the Western world had received since the writings of Malthus, Ricardo and John Stuart Mill in the 18th and 19th centuries. It came when Europe and the USA were on the crest of 25 years of untrammelled growth. So it was severely criticised for being simplistic, anti-poor and insensitive to the full potential of technology. Among its sternest critics were Yale economist Henry Wallich and Robert M. Solow, who went on to win the Nobel Prize for economics in 1987. The burden of their criticism was that 'the authors load their case by letting some things grow exponentially and others not. Population, capital and pollution grow exponentially in all models, but technologies for expanding resources and controlling pollution are permitted to grow, if at all, only in discrete increments'.[1]

The report was also deliberately misrepresented in some quarters to discredit its message. One frequently repeated canard was that it had predicted that the earth's mineral resources would be exhausted before the end of the 20th century.[2] The world had no difficulty, therefore, in forgetting it. But at least two reappraisals between 2000 and 2008, notably one published by the Canberra-based Commonwealth Scientific and Industrial Research Organisation

(CSIRO), have shown that in the 30 years since the Club of Rome published its report, the actual trajectory followed by global economic system has conformed very closely to the trajectory it had sketched out in its 'business-as-usual' scenario.[3]

In the last 15 years, the prognosis has, if anything, grown worse. Population growth has slowed, but not by much. The benefits that science has conferred have far exceeded what the report had anticipated. But these have been dwarfed by the stresses generated by global warming and the approaching exhaustion of fossil fuel and other mineral reserves. The first signs of economic collapse have begun to appear at the impoverished fringes of the global community. Desertification and hunger are unravelling the fabric of society in the African Sahel, where states are failing and economies are being shattered. In them, desperation is pushing the strong towards genocide and terrorism, and the weak into flight. Mass migration in search of security and sustenance has begun, and Europe is feeling its cutting edge. For the relatively affluent nations, the end of the Cold War has brought not peace but a renewed insecurity that has plunged them into an almost constant conflict. Not entirely by coincidence, most of it is taking place in areas that retain the largest remaining reserves of oil and gas.

Humanity's appetite for growth has not, however, diminished. This is not because human beings are innately selfish but because, as the Club of Rome had feared, *the market economy has no inbuilt mechanism that can power down growth without signing its own death warrant.* Such a mechanism will have to be superimposed upon it through concerted human, that is, political, action, just as the welfare state had to be imposed upon capitalism during the heyday of the nation state. But political action can only work when there is a pre-existing state structure dedicated to the stability of the realm. Trade unions were able to force redistributive measures upon industrialised nation states in the 19th and 20th centuries because there was a pre-existent national authority on which they could exert pressure. But there is not even the rudiments of an overarching global authority that can impose any kind of limits upon growth or, more pertinently, direct it away from fossil fuels, today.

A merging of two threats

Today, the threat from the limits to growth is merging with the threat from global warming to push the world towards a systemic breakdown even faster than the most diehard pessimists had anticipated. The threat from air pollution has morphed into a threat from CO_2 concentration. The threat from the exhaustion of mineral resources has sharpened into an increasingly fierce struggle to corner, and regulate their supply well before the remaining reserves in the earth's crust are depleted. Fossil fuels are the focal points of both threats, but we remain trapped in a human ecosystem that cannot do without them.

Depleting reserves are heightening the need for political and military control and therefore steadily increasing the potential for conflict. A 2008 study estimated that the proven, recoverable, reserves of oil and gas that remain would last till 2040–2042 and of coal till 2112.[4] A more recent estimate by British Petroleum (BP), which takes into account the development of fracking and other new methods for extracting hydrocarbons from the earth's crust, has pushed the terminal dates for oil and gas back by 15 to 20 years, but that still takes us to only the middle of this century. Long before that, the 'inflexion point' will be reached at which discovery of new reserves falls behind the annual rise in consumption.[5] That is when the struggle between nations to corner the remaining known reserves will begin in earnest. The potential for violent conflict is therefore also rising exponentially.

A desperate search for solutions

With extreme weather events multiplying, and renewable energy still meeting only 6 per cent of human needs, it is not surprising that the search for a way to limit global warming has acquired an edge of desperation. Most of the proposed strategies centre around one idea: if we can't stop producing GHGs, then let us find ways of limiting the harm they can do to us. One set of stratagems aim at extracting CO_2 from the air. In 2011, Christiana Figueres, the executive Secretary of the UN Framework Convention on Climate Change (UNFCCC), surprised the world by urging it to develop technologies that can suck GHGs out of the air to lower the concentration of CO_2 in the atmosphere.[6] This triggered a flurry of proposals on how that might be done. Doable proposals ranged from (a) reforesting vast portions of the earth's surface, and growing mountains of kelp in the oceans to increase CO_2 absorption, to (b) burying billions of tonnes of compressed CO_2 at least 3 km beneath the sea, to (c) converting agricultural crop stubble into biochar and burying the solid residue, which is mostly carbon, in the ground instead of letting farmers burn it to clear their fields for the next crop.

More imaginative, if not practical, suggestions listed by Australian author Tim Flannery, a former head of the Australian Climate Commission, in a 2015 book, *Atmosphere of Hope: Searching for Solutions to the Climate Crisis*, include seeding beaches and other open ground with crushed olivine, a mineral that absorbs CO_2 rapidly as it weathers; planting quicklime (calcium oxide) or sodium hydroxide in scrubbing towers and artificial 'trees' to react with CO_2 and turn it into chalk or sodium bicarbonate. It has even been suggested that giant chiller boxes be built in the Antarctic that will lower the air temperature by 30°C to make atmospheric CO_2 freeze and fall out as dry ice to be collected and buried as it is formed.[7]

The common feature of all these suggestions is that they are capital- and technology-intensive and horrendously expensive. They can therefore be implemented only by the capital-rich nations, through the global corporate sector. Thus, whether they make a difference or not, all of them will ensure that global economic power remains the monopoly of those who enjoy it today.

Not only will the poorer four-fifths of the world not have any say in these projects but, as the global engineering project described below shows, the poor will also pay most of the price.

It may therefore be a blessing in disguise that these technologies are hopelessly unaffordable, even for the rich nations. Assessing known and projected methods of CO_2 extraction, Flannery has put the cost of CO_2 capture by his methods at \$27–28 per tonne of CO_2. But since 7.8 billion tonnes of CO_2 will have to be sequestered to reduce its concentration in the atmosphere by 1 ppm, bringing the concentration of CO_2 down from the current 403 ppm to the 312 ppm of 1950 will cost close to 19 trillion dollars. But that would be without taking into account the cost of extracting the additional CO_2 that will be generated during the years when this is being done. The current rate of addition is in excess of 2 ppm. So a 20-year programme to reduce the concentration to 312 ppm would have to remove 120–125 ppm of CO_2. It would therefore cost half as much again and require an annual investment of 1.4 trillion dollars. That is a sum of money that the world's governments simply do not have to spare.

There is only one truly promising way of sequestering CO_2: it is to extract it from the air or the sea and convert it into transport fuels. Sea water appears the more promising alternative because it has been absorbing most of the excess CO_2 produced by human beings in the last three decades and now contains 140 times as much per tonne as the atmosphere. Since this is acidifying the sea, warming its upper layers, weakening the thermohaline current, reducing the density of plankton and other nutrients in the surface layers and thereby endangering the very base of the human food chain, decarbonising the sea may be more urgently needed in the long run than decarbonising the air.[8]

The US navy has developed, and is using, a process for doing this on aircraft carriers in mid-ocean. In 2011, this was estimated to cost close to \$6.00 per gallon, but can cost only half as much if carried out on land.[9] The missing link in this proposal is the availability of power. It takes a lot of energy to extract CO_2 from the air or water, but an even greater amount to extract the hydrogen that is needed to synthesise it into transport fuels. Apart from a small amount that is released by oil refineries, the only hydrogen comes from the electrolysis of water. In the simplest of all conversions, 1 tonne of hydrogen can combine with 11 tonnes of CO_2 to yield a total of 12 tonnes of synthetic natural gas, which is an excellent transport fuel and petrochemical feedstock. Since almost 33 MWh of power is needed to produce 1 tonne of hydrogen, the power needed to convert even 1 billion tonnes of CO_2 into synthetic natural gas amounts to 2.75 million MWh (2.75 GWh). Advocates of this route to carbon extraction seldom ask where the extra power will come.

Geo-engineering

If there is no alternative to fossil fuels, and carbon extraction from the atmosphere is unaffordable, the only remaining option is to reflect the sun's rays back into space before they have a chance of reaching the earth. This is the genesis of the most tempting, but dangerous, proposal made so far—to flood the stratosphere, which is a layer of the earth's atmosphere stretching from 10 to 50 km above it, with sulphur dioxide pumped up through a 30 km long pipe suspended from helium balloons, in order to reflect a part of the sun's rays away before they hit the earth. The proposal originates in the discovery, almost three decades ago, that the clouds around Venus are made up of not water vapour as some astronomers had fondly hoped, but of carbon dioxide, sulphur dioxide and droplets of sulphuric acid. The last two are responsible for reflecting 78 per cent of the sun's energy away from the surface of the planet and maintaining a temperature equilibrium, albeit at a hellish 475°C.[10]

The possibility that sulphur dioxide can be made to perform a similar function on earth surfaced when Mount Pinatubo spewed 20 million tonnes of matter, including a very large amount of sulphur compounds, into the stratosphere in 1991. In the year that followed, the average global air temperature fell by 0.5°. Since then the possibility of a similar 'controlled' experiment in geo-engineering has gained the backing of some of the biggest global corporations and foundations in the world.

Its obvious drawback, however, is that the engineers have no control over what else their stratagems will do to the planet. The Pinatubo eruption was followed by a long delay, and overall weakening, of the south-west monsoons over India and a severe drought in the African Sahel the following summer. Other major volcanic eruptions have been followed by similar disturbances of the weather. For the Sahel, South Asia, the American south-west, Australia and other drought-prone parts of the world, the attempt to avert climate change by this route could bring on the very disaster it aims to prevent.[11]

Geo-engineering can therefore bring on the very same calamities—desertification, famine migration and genocide—that it is intended to prevent. In the face of this vast unknown, the mere suggestion that human beings have the capacity to bend nature to their will reeks of an arrogance that has crossed the bounds of reason. What is truly ironic is that it is being advocated by the very same people on the planet who have so far been saying that there is no reason to fear climate change because 'we know too little about the way that nature works'!

Notes and references

[1] This was said by Dr Allen Kneese and Ronald Riker of Resources for the future, available at http://www.businessinsider.com/the-limits-to-growth-and-beyond-2011-12#ixzz3itNgMFiz (accessed on 7 July 2017).

[2] Graham Turner, 'A Comparison of the Limits to Growth with Thirty Years of Reality', *Global Environmental Change* 18, no. 3 (2008, August): 397–411, available at http://www.sciencedirect.com/science/article/pii/S0959378008000435 (accessed on 7 July 2017).

[3] Ibid.

[4] Shahriar Shafee and Erkan Topal, 'When Will Fossil Fuel Reserves Be Diminished?' *Energy Policy*, No. 37. (2009): 181–188, available at https://www.deepdyve.com/lp/elsevier/when-will-fossil-fuel-reserves-be-diminished-LgKoIUPhb6? (accessed on 7 July 2017).

[5] Source: BP Statistical Review of World Energy—2015 Main Indicators, available at https://knoema.com/BPWES2014/bp-statistical-review-of-world-energy-2015-main-indicators (accessed on 10 July 2017).

[6] Fiona Harvey, 'Global Warming Crisis May Mean World Has to Suck Greenhouse Gases from Air', *The Guardian* (5 June 2011).

[7] Tim Flannery, *Atmosphere of Hope: Solutions to the Climate Crisis* (Penguin Books, 2015). See also Tim Flannery, 'A Third Way to Fight Climate Change', *The New York Times* (23 July 2015), available at http://www.nytimes.com/2015/07/24/opinion/a-third-way-to-fight-climate-change.html?_r=0 (accessed on 10 July 2017).

[8] See Chapter 1 for details and references.

[9] Heather D. Willauer, Dennis R. Hardy, Kenneth R. Schultz and Frederick W. Williams, 'The Feasibility and Current Estimated Capital Costs of Producing Jet Fuel at Sea Using Carbon Dioxide and Hydrogen', *AIP Journal of Renewable and Sustainable Energy* 4, no. 033111 (2012), available at http://dx.doi.org/10.1063/1.4719723 (accessed on 10 July 2017), and John Morgan, 'Zero Emission Synfuel from Seawater', available at bravenewclimate.com/2013/01/16/zero-emission-synfuel-from-seawater/ (accessed on 10 July 2017).

[10] Fraser Cain, 'Clouds on Venus' (24 December 2015), available at http://www.universetoday.com/36871/clouds-on-venus/ (accessed on 27 February 2017).

[11] Naomi Klein, *This Changes Everything: Capitalism vs. The Climate.* Chapter 8, pp. 271–276 (Penguin Books 2015).

Chapter Three
Blind Alleys and Red Herrings

In the past four decades there has been a worldwide effort to find, and develop, alternatives to fossil fuels. But this has been hampered at every turn by the reification of the Market Economy, the consequent sole reliance upon the Clean Development Mechanism to bring down carbon emissions and its conversion of what should have been a focussed search for technologies that would permit a complete shift of energy base over a long term into a short-sighted scramble for low hanging fruit. This has led R&D into a succession of blind alleys. This chapter emphasises the need for concerted State guidance, frames criteria for the choice of technologies and appraises the contribution that the currently most popular alternative technologies can make.

Till as recently as 2009, no one paid the slightest bit of attention to renewable energy. Research into as many as 20 different forms of energy generation and carbon mitigation had been going on since the first oil price shock of the 1970s, but it was desultory—considered more a plaything of scientists and engineers in universities than a serious search for alternatives to fossil fuels. Some idea of how little faith anyone had in renewable energy may be had from a report prepared for the G8 summit at Hokkaido in

June 2008 by Tony Blair's Climate Group. This estimated that in 2095, business-as-usual scenarios would cause GHG emissions to rise to 80 billion tonnes of CO_2 equivalent a year. The most this could be brought down, by harnessing every possible alternative energy technology, re-greening large parts of the world and pushing energy efficiency to its limit, was to 52 billion tonnes, which was still almost double the emissions of 2007. Any further reduction of CO_2 emissions, the report concluded, would have to come from carbon capture and storage (CCS) underground and a shift in power generation from coal to gas. 'Renewable energy and nuclear power, that is, all non-fossil fuels together, would reduce emissions by only 12 billion tonnes', the report concluded.

An assessment by UN Secretary General Ban Ki-Moon in 2012 was only slightly more optimistic: the share of renewables, which stood at a paltry 6 per cent in that year could, he believed, be increased to 12 per cent by 2033. Neither assessment showed the least glimmer of doubt that there was no alternative to fossil fuels in the foreseeable future.[1] This explains why former president of the United States Barack Obama, whose commitment to limiting global warming was sincere, nevertheless allowed Royal Dutch Shell to resume drilling for oil in the Arctic and defended Canada's extraction of oil from Bitumen in Alberta during the visit of Canadian Prime Minister Justin Trudeau to Washington in November 2015 only weeks before the start of the Paris summit. For the absence of alternatives left him with no choice. Every other government has been facing the same dilemma.

What is less easy to understand is why, despite being fully aware, even in 2008, of the risk of annihilation that the world was running, the G8 governments had done so little to avert it, and why they continue to do so little today. Had their professions of concern for the future of the planet been sincere should they not have been pouring money into the research and development of renewable energy alternatives? It was not as if these technologies were completely unknown: solar thermal energy had been providing peak power in California, since 1988, and the EU had certified a Swedish technology for converting black liquor (a noxious and poisonous effluent of the paper industry), into methanol for use as a fuel additive in place of ethanol, two years before the Hokkaido summit. Solar photovoltaic (SPV) and wind power were already in commercial use.

But instead of pouring money into the further development of these technologies, they did, and continue to do, the exact opposite—pour funds into fracking, dish out permissions to hunt for new reserves of oil and gas beneath Greenland's glaciers and under the Arctic, resist calls to impose a global carbon tax and continue to spend $550 billion a year on subsidising the production and consumption of fossil fuels, while devoting only $6 billion, a paltry 2 per cent, of their overall research and development expenditure, on the development of renewable energy.[2]

The explanation of this anomaly does not lie in ignorance. It does not even lie in the undoubted power of the fossil fuel lobby. It lies in the enslavement of the human mind to a specific ideology—an enslavement that is so complete that few people are conscious of it. This is the ideology of the Free Market. The market economy exalts the virtues of competition, because competition ensures that only the best succeed: the best products, the best managers and the best technologies. The Free Market therefore maximises production and wealth and rewards the deserving while 'weeding out' the undeserving. It is therefore the economic counterpart of Darwin's theory of the Origin of Species. And like Darwin's theory, it is a good substitute for God.

Applied to the realm of energy, the sanctity of the Free Market virtually mandates leaving the choice of technologies that will succeed fossil fuels to the market. Since profit is the yardstick of success, reliance upon the market will automatically lead to the choice of the lowest cost sources of energy.

The market economy does not deny a role to the government in the choice of technologies, but dictates that this should be minimal and should under no circumstances impede the free interplay of market forces. In effect, this means that the governments can set targets and offer inducements to meet them but 'it must offer them to the market as a whole and not to individual firms or technologies'. It explains why in meeting after meeting on climate change, the representatives of industrialised countries never discussed the advantages and limitations of specific technologies, their respective capacities to replace fossil fuels, and their limitations. All that was the business of 'the market', and therefore the preserve of private enterprise. It was therefore a no-go area for governments.

Clean development mechanism— The invisible hand rises again

This unquestioning faith in the market mechanism led directly to the Kyoto Protocol's exclusive dependence upon the Clean Development Mechanism (CDM) to bring down carbon emissions. The CDM awards Certified Emission Reductions (CERs) certificates, dubbed carbon credits, to companies that introduced new, or adopted existing, carbon-reducing technology. Each CER denotes a reduction of 1 tonne of CO_2 emissions. The recipients were allowed to sell these to other companies that had fallen short of their targeted emission reduction in a given year. In this way, innovators earned rewards by selling their CERs, while laggards were penalised by being forced to devote a part of their profits to the purchase of CERs. The first carbon credits were issued under it in 2001, but the programme gained momentum only after the Kyoto Protocol came into force in 2005. By early 2016, 1.6 billion CERs had been awarded to 8,000 firms.[3] In absolute terms, this was not a small achievement, but it barely made a scratch on the growth of CO_2 emissions which grew by a billion (giga) tonnes a year from 2001 to 2015.

On paper, the CDM is a liberal economist's dream come true: it costs the state nothing; it keeps the state out of the market and it maximises efficiency by allowing 'the free interplay of market forces' to determine which technologies will succeed and which will fail. But that is where the problems arise. For the market has only one yardstick of success, profit, and that too the profit of the investor, not of society. In recent years even this has been transmogrified into 'anticipated' profit, judged every quarter by armies of market 'analysts' whose predictions move the share market. The dubious nature of this yardstick of success can be judged from the number of times it has been manipulated to yield desired assessments to privileged customers in recent years. It is to this inherently flawed and corrupted institution that the CDM has ceded that responsibility of saving the planet.

In the literature on climate change, the CDM's failure has been ascribed to a wide variety of causes, such as excessive red tape, long delays in the award of CERs, expensive procedures for application which smaller companies cannot afford and the exclusion of agriculture.[4] But the CDM would have failed even if the award of CERs had been instantaneous and the application process cost-free, because the blind faith upon which it is based—that a billion separate actors pursuing their own profit oblivious of what the others are doing can deliver a benefit to society that they are not even looking for and will take at least half-century to emerge— is backed by neither reason nor experience. This is the 'invisible hand' that has grown so large that it now holds the globe in its fist.

What the CDM has done is deprive the search for solutions of any sense of direction. It has sent companies running helter-skelter in every direction looking for quick ways to earn carbon credits. So it is only by accident that any money has gone into technologies that hold a serious promise of shifting mankind's energy base. For instance, 57 per cent of all the CERs issued till the end of 2012 went to companies that had eliminated hydrofluorocarbon (HFC) refrigerant gases and nitrous oxide, which reduced other GHGs but did nothing to replace fossil fuels.

The three most frenetically developed renewable energy technologies, wind and PV electricity, and ethanol from sugar and corn, are all products of the hunt for quick profits. For all require relatively little capital and come on stream in as little as a year. So while all three can shower carbon credits upon their investors, none can provide base-load power, or the sheer volumes of transport fuels needed to do what fossil fuels are doing today.

Market forces can improve the functioning of an economy working on a given energy base but cannot by themselves change, much less decide in which direction to change, the base itself. In the era of the Nation state, this could have been done by the state. But in the globalised, post-national, world it can only be done by a concert of nations. In either case, the principle remains the same: decisions that affect society must be taken by an organisation that it can hold accountable—in other words a national government or a global quasi-state authority. By inserting the market between state and society, the CDM has made this impossible.

This realisation has sunk in. On 15 June 2015, under the auspices of the Grantham Research Institute on Climate Change and the Environment of the London School of Economics, seven eminent policy makers in Britain launched 'A Global Apollo Programme to Combat Climate Change', the purpose of which is to 'radically cut the world's output of carbon dioxide'. Quoting Thomas Edison's remark, 'I'd put my money on the sun and solar energy', they said that only tapping solar energy would make this possible. The world, therefore, needed a programme similar to the US' Apollo Programme of the 1960s to put a man on the moon that would be well funded and bring the best scientific and engineering talent together, to make clean energy less costly to produce than energy based upon coal, gas and oil. To do this, they said, 'We need a major international scientific and technological effort, funded by both public and private money'. The target they proposed was to make 'new-build base-load energy from renewable sources become cheaper than new-build coal in sunny parts of the world by 2020, and worldwide from 2025'.

These target dates, which are based upon Bloomberg's estimate of the levelised cost of electricity from 20 renewable energy sources in the first half of 2014, have turned out to be conservative, for contracts for SPV and concentrated solar thermal energy began to be signed in places such as India and Morocco, at prices fully competitive with modern, environmentally safe, coal plants from the end of 2015.[5]

But the Grantham Research Institute's proposal has served to bring economics and technology into a discussion that has so far been dominated by diplomats and trade specialists. It has also warned policy makers that henceforth it is the state that must lead the search for an alternative energy base, instead of playing handmaiden to the market. This change of roles is long overdue.

The role of the state

What are the criteria upon which the world's governments should decide which technologies to promote? One way to answer this question is to pose a counterfactual one: suppose scientists came forward and told governments, 'We have found a way of producing energy free of cost. All we want you to do is convince yourselves that it works and set a date after which no one will be allowed to use fossil fuels'. Would that be the end of the world's problems? The answer is 'no'. For the world does not need just 'free' energy: it is getting all that it needs, and more from the sun. It needs the right amount of energy, in the right concentration, at the right place, the right time, in the right form and with the required degree of reliability. To ensure this, it has built an elaborate infrastructure of roads, railways, sidings, power stations, transmission grids, oil and gas pipelines, terminals, refineries and gas stations over the last 200 years. If the new, zero-cost energy can be delivered through the same infrastructure, it will be genuinely cost free. But if it requires, even in part, the construction of an entirely new infrastructure, all that collateral cost will have to be recovered from the users. Then it will be free no longer. 'Free' energy is therefore an oxymoron.

One reason why fossil fuels retain their dominance even when their reserves are approaching exhaustion is that the energy delivery infrastructure they need has already been built and fully depreciated. At the margin, therefore, its additional cost of supply *to society* is close to zero. An important goal of state intervention should therefore be to identify and promote technologies that promise not only to be profitable to the investors but to impose the lowest cost of adjustment upon society.

In sum, these have to:

1. be technologically proven or close to being proven;
2. be profitable in terms of private costs and returns to the investor at current market prices, or close to being so;
3. be available as and when in the quantities required;
4. minimise external social costs by causing as little disruption of the existing production, distribution, marketing and pricing infrastructure as possible;
5. alter the existing distribution of income and assets as little as possible;
6. make the shift as seamless as possible by posing the minimum threat to investors who have invested heavily in existing energy technologies.

When the available renewable energy options are measured against these yardsticks, our choices get narrowed considerably.

Wind power

Wind power is the first, and so far the most fully exploited, source of renewable energy to have been harnessed. In terms of the raw energy it contains, wind comes second only to direct solar energy. Windmills were the first external, non-biological, source of energy harnessed by human beings. This led to the first great surge of productivity in Europe in the 10th and 11th centuries that sowed the seeds of its rise to global pre-eminence eight hundred years later. So it is understandable that the possibility of harnessing wind power once again has a special fascination for European governments today. Since 2000 AD, wind power generating capacity has been doubling every three and a half years to reach 12 times the 2000 figure in 2014.[6] Wind farms are now providing around 4 per cent of the electricity the world consumes. But its advocates see no limit to what it can do. In its publication *Technology Road Map: Wind Energy 2013*, the International Energy Agency (IEA) predicted that a tenfold increase in its 2012 generating capacity of 282,000 MW (0.28 TW) to around 2.8 million MW (2.8 TW), coupled with further advances in technology, would enable wind to meet 18 per cent of the world's demand for electricity in 2050 AD. This might be technically feasible, but with wind generation requiring a quarter of a square kilometre of land per MW—a ratio unlikely to change by much in the future—this would require a minimum of 700,000 km^2 of the earth's surface. Without going far offshore this would be almost impossible to find.[7]

There is also a widespread misconception that the availability of wind energy is virtually limitless. In a study published in 2009, three members of the engineering faculty at Harvard University estimated that harnessing wind energy alone could produce 40 times as much electricity and five times as much energy, in all its forms, as the world consumes today.[8] The study also pointed out that the two principal emitters of CO_2, China and the United States, had the maximum wind energy potential—18 and 23 times, respectively—in relation to their 2008 consumption of electricity. Two years later, another study of the feasibility of replacing fossil fuels with wind, water and solar power concluded that the world's estimated demand for electricity in 2030 could be fully met by tapping less than 1 per cent of the energy contained in the wind.[9]

But in their enthusiasm, not only did both sets of authors grossly overestimated the amount of raw wind energy actually available for conversion into power but also overlooked the fact that its conversion into usable power is limited by the second law of thermodynamics. This is that no transfer of energy can take place without a rise in entropy, that is, without some loss of the original energy. The conversion of raw wind energy into usable electricity goes through so many stages of conversion that the amount of usable power that remains is a fraction of the original.

The total kinetic energy in the winds that swirl in the earth's atmosphere is indeed many, many times what would be required to meet mankind's demand for electricity. But most of it is to be found in the upper layers of the atmosphere, notably in the jet stream. Only a fraction of it filters down to the 200 m at the earth's surface that the largest windmills under development today (8–10 MW) will occupy.

Carlos de Castro, professor of applied physics at the University of Valladolid, in Spain, estimated electricity that can be gener-ated from wind in a paper titled 'Global Wind Power Potential: Physical and Technological Limits', in 2011.[10] Pointing out that estimates of the total kinetic energy generated by the sun in the form of wind varied widely from 340 to 3,600 TW (million MW), de Castro based his calculations on a middling figure of 1,200 TW. Of this, only one-twelfth is available in the layer of air up to 200 m from the earth's surface. This means that the usable kinetic energy is only 100 TW. But four-fifths of the earth's surface is covered by deep seas, human habitation, permanent ice and forests. This reduces the available kinetic energy on non-urban, easily accessible land and shallow coastal waters to 20 TW.

The need to space out the windmills further reduces the energy that reaches the blades, to about 30 per cent of this figure, that is, 6 TW. But to generate power, the winds must also be neither too slow nor too fast for the windmills to withstand. Castro assumes that suitable winds will be available for half the hours in the year. This brings down the available kinetic energy to 3 TW.[11]

But there are still two more steps to go: first, the blades have to overcome the inertia of the turbines; second, the mechanical energy they impart to the turbines has to be converted into elec-tricity. In 2009, the efficiency of transfer of mechanical energy from blade to turbine was about 50 per cent, but was expected to rise to 75 per cent in the near future. Castro, therefore, assumed 75 per cent. Second, the theoretical, but not yet achieved, limit of turbine efficiency is 59 per cent, the current best being around 40 per cent. So Castro and his co-authors based their estimates on a turbine efficiency of 50 per cent. On this basis, they concluded that the electricity-generating capacity of winds within 200 m of the earth's surface is probably not more than 1.125 TW a year of continuous generating power.[12] In sum, if Castro's assumption that the total kinetic energy in global winds is 1,200 TW, and that suitable winds are available for half the year is correct, the IEA's estimate of a generating capacity of 2.8 TW is more than double of what is available in the bottom 200 m of the atmosphere.

Apart from the insufficiency of raw kinetic energy, wind power has the insurmountable drawback of being available only intermittently. Castro's second assumption that suitable winds are available for half the year is true of only a few locations, mainly in northern Europe, North America and China. In the world as a whole suitable winds are available for 21 per cent of the year. So if Castro has not underestimated the kinetic energy in the wind envelope of the earth, then 0.6 TW is the most that wind farms will be able to provide without building platforms far offshore in the deep oceans. Wind power cannot, therefore, give humanity the control over its sources of energy that modern life requires.

Solar photovoltaic power

Generating electricity directly from the sun has been the Holy Grail of environmentalists. In 1918, a Polish scientist Jan Czochralski developed a method for growing large single crystals of metal. This was adapted 30 years later by an American scientist Gordon Teal to produce large, extremely pure crystals of germanium, for use in transistors. Six years later, Daryl Chapin, Calvin Fuller and Gerald Pearson of the Bell Laboratories adapted Teal's method to develop the first silicon PV solar cell capable of converting enough of the sun's energy into power to run everyday electrical equipment.

In 1932, two other scientists Audobert and Stora had discovered the PV properties of cadmium selenide. This opened the way for the development of thin-film PV cells that deposit a microscopically thin coating of chemicals such as cadmium telluride and gallium arsenide onto plastic to create cheap, light and flexible PV panels. Half-century later, these two discoveries fused to create the PV revolution of the 1990s. The cost of a PV cell, which was US$76 in 1977 when made of crystalline silicon, fell to 77 cents in 2013 and has fallen further since then.[13]

By the beginning of the 2000s, a silent revolution had begun in SPV power generation. SPV 'modules'—complete systems—for supplying power directly to the grid began to be mass produced. SPV was also the main beneficiary of the wave of fear ignited by the Fukushima Nuclear Power Plant disaster in March 2011. Initially the decision to shun nuclear power also brought the high cost of renewable energy alternatives into the limelight. In Germany, which took the lead in replacing nuclear power with renewables, the government was, till 2013, paying a subsidy of 0.17 euros (24 US cents) per unit on electricity from renewable sources.

But in 2014, exports from China, which was producing half of the world's solar panels, and a sharp drop in world raw material prices, brought the cost of PV power generation down to a level that was competitive with, and even below, conventional power. This made further subsidisation unnecessary.[14]

Clinching proof that SPV energy had become fully competitive with coal came in 2015 and 2016. Four giant SPV power stations were auctioned in India between November 2015 and January 2016, at power purchase prices of 6.00 to 6.4 US cents (₹4.34 to ₹4.63) per unit. This was fully competitive with coal.[15] But SPV costs continued to fall, for in August 2016 a contract was awarded in Chile for a power plant that would charge only 2.91 cents a unit or half the cost of electricity from coal.[16] This enormous drop in contracted price, which took the SPV industry by surprise was not the product of another dramatic breakthrough in raising panel efficiency, for in March 2014 First Solar, the US-based pioneer in SPV generation, announced that, having achieved a full cell area conversion efficiency of 17 per cent it hoped to raise it only to 18.9 per cent by 2017. The Chilean contract was almost entirely the result of innovative finance designed to bring down the cost of capital, and of the decline in raw material costs.[17]

SPV too cannot provide base-load power

Since more than two-thirds of the land area of the world is either desert or semi-arid, there is no limitation of physical space for SPV plants. But SPV too has the crippling drawback that it cannot generate power when the sun is not shining. In Paris, on an average, the available sunshine for SPV generation is only 3.34 hours per day, giving a maximum plant utilisation of 14 per cent. As Figure 3.1 shows, the average availability of SPV plants between 45° north and south of the equator is about 16 per cent of the year. As a result, despite having two-thirds of the installed capacity of wind power, SPV was meeting only 1 per cent of global electricity consumption in 2014.

SPV has a second drawback that its promoters almost never mention: the efficiency of solar panels decreases by half a per cent for every degree Celsius rise in temperature above ambient room temperatures. In countries with extreme climates, where the temperature in the sun during peak daylight hours in summer can average 80°C, this can halve the efficiency of the panels if they are not cooled. Simple cooling systems can be installed but will increase costs and use up some of the energy that SPV generates.

Finally SPV power also suffers from the problem that limits the contribution that wind power can make towards shifting the world out of fossil fuels: there simply is not enough of rare earths (REs) and heavy metals available in the earth's crust to make more than a small dent in the empire of coal. This is discussed in greater detail in the next chapter.

Figure 3.1: Load factor versus latitude, 419 solar PV systems

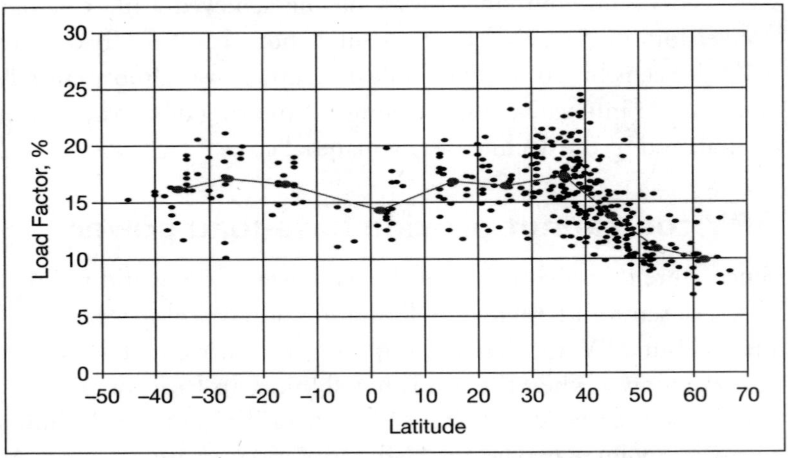

Source: Roger Andrews, 'Estimating Global Solar PV Load Factors' (20 June 2014), available at http://oi61.tinypic.com/2rrt5iw.jpg (accessed on 12 July 2017).

Power from tides

Tidal power suffers from many of the same problems as wind, although the 'World Offshore Renewable Energy Report 2002– 2007', released by the UK Department of Trade and Industry, esti- mated the potential for tidal energy at a staggering 3,000 GW, less than 3 per cent of this potential is located in areas suitable for power generation. This is because the tidal current is strong enough to generate electricity locations, where its amplitude is increased by specific features of local topography such as the rate of shelving of the sea bottom, funnelling in estuaries and reflection by large peninsulas.[18]

Of 3,000 GW, 3 per cent, that is, 90 GW, is still a large amount of power. But as in the case of wind energy, the propor- tion of the kinetic energy present in the water that can actually be converted into electricity will be a good deal less. Add to this the fact that it is a discontinuous and fluctuating form of energy that will wax and wane with the speed of flow of the water and stop altogether when the tide turns, that the tides are strongest in remote and sparsely populated regions from where the electricity will have to be transported long distances and it becomes apparent that while tidal power can meet a part of the energy needs of local coastal communities located in the cool temperate regions, its contribution to limiting climate change will be negligible.

Geothermal energy

Geothermal energy, trapped in naturally occurring hot spots in the earth's crust, has one immense advantage over wind and tides—it can generate electricity continuously and can therefore directly replace both fossil and nuclear energy for power generation. At the end of 2012, there were plants with 10,715 MW of power-generating capacity online in 24 countries, with the United States far ahead of the rest at 2,800 MW.[19] But all of it was of the conventional kind, located close to tectonic plate fault lines and was generated by tapping natural hotspots where water was trapped as steam.[20]

However, the potential for tapping this abundant and totally clean form of energy has barely been tapped, for the technology that has been developed for extracting natural gas and 'tight' oil from shale beds tar sand beds deep underground, can also be harnessed to tap heat trapped much further down within the earth's crust. This process has been labelled an *Enhanced Geothermal System* (EGS). Installing an EGS plant typically involves drilling a 25 to 30 cm wide hole 3 to 4 km down into the earth's crust, expanding existing fractures in the rock at the bottom of the hole by pumping down water under high pressure and drilling a second hole into those fractures. Water pumped down one hole then percolates through the cracks, turns into steam and flows back to the surface through the second hole where it runs a steam turbine, provides process heat and cools down. The re-condensed water is then sent back to the hot zone through the first hole.

EGS has been studied extensively in the United States since 1977 and was taken up in Australia and the EU in the early 2000s.

But it has not taken off so far. Although an MIT study in 2006 estimated that EGS systems could deliver power in the United States for as little as 3.9 cents a unit, the technology has not so far attained commercial viability.[21] The onset of recession after 2007 and a diversion of available resources to fracking could explain this. But it is also possible that the MIT study was too optimistic. For the viability of EGS power depends critically upon the temperature of the rock core into which the water is injected and the amount of water available. The former has ranged from 150°C to 200°C depending upon the depth at which the heat is tapped. This is low for steam turbine power generation which is now taking place at close to 500°C. It, therefore, yields a correspondingly lower turbine efficiency.

EGS also requires very large volumes of water to pass through the rock, typically a thousand gallons (4,000 litres) per MW per minute. This is abundantly available as associated water in onshore oil and gas fields, but depending upon this alone would make EGS a captive to the continued production of oil.[22] Elsewhere, the availability of water, which is bound to decline in the coming years, as stated in Geothermal Energy Association's report, will limit the adoption of EGS power systems.[23]

On the brighter side, geothermal energy is being extracted from conventional systems for close to 6,500 hours a year, which amounts to a plant utilisation of close to 75 per cent.[24] It is therefore entirely capable of meeting base-load demand for electricity. EGS has been able to do this with the technologies that exist today in only a few places, but focused research on developing them could make it a major source of base-load electricity and process heat in the years to come.

The lure of over-the-horizon technologies

An over-reliance on the CDM has not only vitiated research on technologies that have the capacity to meet our energy needs, but vigorously promoted a search for over-the-horizon technologies that will take so long to harness that they will make the world run out of time. Hardly a year passes without reports of breakthroughs in 'green' technologies for combating global warming. These share several common characteristics: they involve dramatic departures from existing technologies; they are at the laboratory or at most pilot demonstration stage of development and will therefore take decades to commercialise and they are all ruinously expensive. The promoters, who are very often the companies that are most heavily invested in the current generation of fossil fuel technologies, assure us that large-scale production will one day make these technologies economically viable. In the meantime, in an ironic reversal of the infant industry argument, they demand heavy subsidies during the scaling-up period. Three recent, much hyped, 'breakthroughs' illustrate how this works.

Hydrogen-fuelled automobiles

In June 2008, the Japanese car company Honda unveiled the world's first commercially available, totally environment-friendly car, named the FCX Clarity.[25] Propelled by the long-awaited hydrogen fuel cell, it was a family saloon capable of accelerating from zero to 100 km/hr in 9.8 seconds and of running more than 400 km without needing a recharge of its fuel cells. Honda announced that it intended to lease a limited number of prototypes to selected customers at a mere $600 a month—which was about the lease price of a Porsche. Should the world have breathed a sigh of relief? Had technology provided the perfect zero CO_2 antidote to the emission of CO_2 in the transport sector? A closer examination of Honda's plans showed that its plans for a hydrogen-fuelled car had relatively little to do with global warming and a lot to do with warming the bottom line of the company.

Honda did not release the cost of its first hydrogen-powered cars because it intended to produce only 26 of these cars and provide them to a few, very high profile, celebrities in order to gain maximum exposure for their new technology. But the $600 monthly lease rate at which it released the car the FCV Clarity two years later suggests that the prototypes had cost a great deal more than $100,000 that the US Department of Energy had estimated six years earlier. In 2008, the company stated that it expected to bring down the cost to $100,000, but that this would take several years. By then it expected the rising cost of gasoline to make it an attractive, not to say novel, alternative to the Porsche or Ferrari. By 2010, Honda had placed only 20 such cars with celebrities. In 2014, it announced the termination of the Clarity project. By then it had leased, in all, 45 of these cars in the United States, 10 in Europe and another 10 in Japan.[26]

The lure of hydrogen-powered cars, however, has remained alive. Chevrolet, Hyundai, Mercedes and Toyota are the more recent entrants into the field. But their cost remains prohibitive. The Toyota Mirai has been launched in Japan at $70,000. It is a family saloon with a claimed range of 310 miles as opposed to 190–231 miles for the Clarity and other makes,[27] but according to the online magazine *CleanTechnica*, Toyota will be losing $100,000 per car, at least till the scale of production increases dramatically.[28] However, this is not likely to happen soon. For the hydrogen that is competitive with gasoline prices—currently estimated to cost $3.00 a gallon in the United States—is actually being produced from natural gas and is therefore anything but 'green'. It is a highly efficient fuel but does not allow rapid acceleration, and the distance it can travel on a full tank is limited. There are very few hydrogen-filling stations, and these are concentrated almost entirely in California and New York. Both the cost of the car and its lukewarm reception by consumers is discouraging gas stations from making the necessary investment in a new infrastructure. Until a way can be found to produce hydrogen cheaply from renewable sources, hydrogen cars will polishing up the makers' corporate image, but will have as much to do with reducing global warming as a Cartier necklace or a Ferragamo handbag.

Carbon capture and storage

Two other much-touted technologies have also flattered only to deceive. The first is the much discussed CCS—burying the carbon dioxide that is generated by coal-burning power plants under high pressure, in natural faults and pockets deep underground. The second is the replacement of gasoline with ethanol. Capturing carbon dioxide from power station emissions is easy. Storing it underground on any significant scale is not only difficult but expensive as well. In 2007, a German company Vattenfall Europe AG began the construction of a 120-MW 'zero-emission', coal-based, demonstration plant that would capture 9 tonnes of CO_2 per hour or 85,000 tonnes a year. The promoters, who were promptly awarded a large subsidy by the EU, intended to store it deeply and securely underground and ruled out the possibility of its ever escaping into the atmosphere. The example set by Vattenfall took a world that craved 'quick fixes' by storm. By April 2009, the United States and various governments in the EU had approved 238 CCS projects, saying they were willing to subsidise them. Three-fifths of these were at the research, pilot or demonstration stage, but a hundred were to be commercial-scale plants.[29] But six years after the pioneering effort at Vattenfall, CCS has been rejected all over the world in all but name. Estimates made at the National Energy Technology Laboratory (NETL) and MIT in the United States had shown that extracting pure CO_2 from the flu gas, compressing it, pumping it kilometres underground and sealing the reservoirs effectively would add between 75 and 90 per cent to the capital cost of older plants and 60 to 85 per cent to the cost of newer plants that worked at supercritical temperatures. In 2007, the fourth IPCC report had also estimated that the cost of electricity from a coal-based carbon capture plant would be double the cost of power from a natural-gas-based combined cycle plant. But this was only an estimate, as no such commercial-scale plant has been built so far.

Subsequent studies gave more precise numbers to the cost of CCS. Among the approved projects, those with known costs were estimated to require an investment of $44.5 billion to sequester 60 million tonnes of CO_2 per year. This amounted to $741 per tonne of CO_2—about 12 times the highest price that carbon credits had commanded in the European market. To capture all of the energy-related CO_2 emissions will, therefore, require an investment of $22.25 trillion, and all of this would have to be justified on the grounds of its social benefits, for it would not yield even one additional dollar of revenue to the power plant.

Cost is the most important but not the only reason why CCS has failed to take off. No sooner had the first projects been announced than it became clear that no one wanted a CCS burial site in their backyard, because no one could guarantee that the CO_2 would remain buried forever. But big business was not prepared to give in. The solution, it says, is to store the CO_2 beneath the sea at depths of up to three miles. The huge pressure of the ocean above, they claim, will make sure that the CO_2 that is pumped down will stay down. If finance were no hurdle this might work, but the reaction of the earth to such an invasion is poorly understood. Pumping CO_2 into its cracks and fissures could be riddled with side effects that no one can foresee. As with pumping sulphur dioxide into the atmosphere there is no scope for laboratory testing such an expedient. CCS will be a blind leap into the dark with fingers crossed.

What little we know of the earth's past behaviour is not reassuring. Earthquakes are known to have broken clathrates and release vast amounts of CO_2 into the atmosphere in the paleo-climatic past. The quantities that will have to be stored are in any case mind-boggling. In his book *The Revenge of Gaia*, James Lovelock estimated that, if frozen, the carbon dioxide we were producing at the start of this millennium would create a mountain a mile high and 12 miles in circumference *every year*. Since then annual emissions have increased by half as much again. Compressing this amount and releasing it into underground structures, the properties of which we do not know, will amount to hanging a sword of Damocles one's own head and waiting for it to fall.

CCS—The technology so far avoided

Storage underground a huge and expensive red herring. What is surprising therefore is the eagerness of the United States and EU to sink up to $50 billion into this experiment, when they have not put a single penny into its tried and tested alternative—to capture flue gas CO_2 by turning it directly into methanol, natural gas or other transport fuel. This process came briefly into commercial use in 1922, but was edged out by the associated natural gas piped from oil fields. There is one power plant doing this even today, at Coleson Cove in New Brunswick, Canada, but no company or government, least of all the Canadian government, has come forward to learn from its experience.[30]

Cellulosic ethanol—A red herring

The third red herring is cellulosic ethanol. The US Energy Independence and Security Act of 2007 set a production target of 16 billion gallons of ethanol. But the administration already knew that this would barely make a dent in the consumption of gasoline, for in 2006, Alan Greenspan, the former chairman of the US Federal Reserve Board, had told the US Senate Committee on Foreign Relations that he expected another over-the-horizon technology to solve the problem. This was the production of cellulosic ethanol from crop residues, grasses and other types of biomass.

In the scramble that followed Greenspan's observation, dozens of companies leaped into the fray with claims that they had perfected enzymes that could speed up and increase the extraction of alcohol from cellulosic biomass through fermentation. In 2007, therefore, the Bush administration allocated $385 billion of grants to seven American companies to produce cellulosic ethanol. But nine years later, all of these had failed. Most research was still at the pilot plant stage. Only one technology, based on a single biomass feedstock, had been developed to a commercial level. That was as an adjunct to a sugar mill at Piracicaba, Brazil, which came on stream in August 2015. This plant is converting bagasse into ethanol and is expected to raise its output to a billion litres (approximately 750,000 tonnes) a year in 2024. Bagasse makes up only one quarter of the total sugar cane crop wastes that can be turned into transport fuels. And it is only one of a score or more of important crop wastes, each of which will fight the enzymes that seek to dissolve it, in its own way. Completing the research needed to find the right enzyme for the right crop waste will take decades. Despite that, cellulosic ethanol is still firmly listed as a 'second generation' biofuel that will come to humanity's aid within the next decade, and millions of dollars of research grants continue to be poured into ongoing efforts to find new enzymes and stabilise its production from different types of biomass.

Given the speed at which the next energy shift must be made if climate change is to be minimised, the fermentation route to producing transport fuels from biomass is a non-starter and should be actively discouraged. This is because another way of doing so, via thermo-chemical conversion, is not only well known, but been in sporadic commercial use in the United States in the last century, and in Canada, and Sweden in recent years. Only a carefully constructed denial of knowledge permits the gadarene rush taking place today down a road that leads nowhere. Why this is happening, and who, if anyone, might be gaining from this is dealt with later in the book.

A desperate search for solutions

So far there is no source of energy that can replace coal and gas as sources of base-load electricity and process heat. So it is not surprising that the hunt for ways to limit global warming has acquired an edge of desperation. More and more of the recent proposals to arrest the heating of the planet centre around the idea that if we cannot stop producing GHGs then let us at least extract it from the air even as we produce it, till we find a better solution. In her book *Climate Wars*, Gwynne Dyer concedes, with more than a tinge of regret, that this will have to be one of four stratagems that mankind will have to resort to if it wishes to avoid conflicts over food, water and mineral resources as the earth desertifies.[31] In 2011, Christiana Figueres, the executive secretary of the UNFCCC, surprised the world by urging it to develop technologies that can suck GHGs out of the air to lower the concentration of CO_2 in the atmosphere.[32] This triggered a flurry of proposals on how that might be done. Doable proposals ranged from (a) reforesting vast portions of the earth's surface and growing mountains of kelp in the oceans to increase CO_2 absorption, to (b) burying billions of tonnes of compressed CO_2 3 km beneath the sea, to (c) converting agricultural crop stubble into biochar and burying the solid residue, which is mostly carbon, in the ground instead of letting farmers burn it to clear their fields for the next crop.

More imaginative suggestions included seeding beaches and other open ground with crushed olivine, a stone that absorbs CO_2 rapidly as it weathers; planting quicklime (calcium oxide) or sodium hydroxide in scrubbing towers and artificial 'trees' to react with CO_2 and turn it into chalk or sodium bicarbonate. It has even been suggested that giant chiller boxes be built in the Antarctic that will lower the air temperature by 30°C to make atmospheric CO_2 freeze and fall out as dry ice to be collected and buried as it is formed.[33]

Perhaps the most promising of these is to extract CO_2 from the sea and convert it into transport fuels, thereby creating 'space' in the sea to absorb more CO_2 from the air. Sea water is currently absorbing most of the excess CO_2 being produced by human beings and now contains 140 times as much CO_2 per tonne as the atmosphere. The US navy has developed, and is using, a process for doing this on aircraft carriers in mid-ocean that in 2011 was estimated to cost close to $6.00 per gallon, but would cost only half as much if carried out on land.[34]

Assessing known and projected methods of CO_2 extraction, Tim Flannery, author of a forthcoming book *Atmosphere of Hope: Searching for Solutions to the Climate Crisis* has put the cost even lower at $27–28 per tonne of CO_2. But since 7.8 billion tonnes of CO_2 would need to be sequestered to reduce its concentration in the atmosphere by 1 ppm, even at this low cost, bringing the concentration of CO_2 down from the current 400 ppm to the 312 ppm of 1950 will cost close to 19 trillion dollars. And that is without taking into account the cost of extracting the additional CO_2 that will be generated while this is being done. This is money that the world's governments simply do not have.

No one knows how many of these proposals will prove feasible. But we do know that, in a world where the highly industrialised countries face long-term deindustrialisation, chronic unemployment and rising social welfare costs, but continue to sequester more and more of the available funds for fighting wars and terrorism, these will prove unaffordable. Assessing a mix of known and projected methods, Tim Flannery, author of a recent book on carbon extraction from the air,[35] has estimated that it could cost $100 to sequester a tonne of carbon (or $27–28 per tonnes of CO_2).[36] But since 7.8 billion tonnes of CO_2 (2.133 billion tonnes of carbon) would need to be sequestered to reduce its concentration in the atmosphere by 1 ppm, even at this low cost, bringing the concentration of CO_2 down from the current 400 ppm to the 312 ppm of 1950 will cost close to 19 trillion dollars. And that is without taking into account the cost of extracting the additional CO_2 that will be generated while this is being done. This is money that the world's governments simply do not have.

In any case, carbon extraction from the atmosphere is a dead end. For scientists at the Carnegie Institution for Science, in the United States, have estimated that even if all the CO_2 added since 1900 is removed from the atmosphere at one stroke and current anthropogenic emissions reduced to zero, it will only bring down mean air temperatures by a little less than half of the rise that has taken place since then. This is because much of the 'space' left in the atmosphere will be filled within five years by a reduced exhalation of CO_2 by plant life, as it struggles to conserve carbon for its own growth and the release by the sea of a part of the CO_2 it has been forced to absorb in recent decades. To ensure that the respite is not temporary, they conclude, there will have to be a 'continuous removal of anthropogenic carbon dioxide' for 80 years thereafter.[37]

Geo-engineering

If there is no alternative to fossil fuels, and carbon extraction from the atmosphere is both unaffordable and ineffective, the only remaining option is to reduce air temperatures directly. This is the genesis of the most tempting, but most dangerous, proposal made so far—to 'geo-engineer' a succession of mild non-nuclear winters by flooding the stratosphere, a layer of atmosphere stretching from 10 to 50 km above the earth, with sulphur dioxide pumped up through a 30 km long pipe suspended from helium balloons, in order to reflect a part of the sun's rays away before they hit the earth. The proposal is rooted in the observation that when Mount Pinatubo spewed 20 million tonnes of sulphur compounds into the stratosphere in 1991, the average global air temperature fell by 0.5° in the following year. Since then geo-engineering has gained the backing of some of the biggest global corporations and foundations in the world.

Its obvious drawback, however, is that the Pinatubo eruption was also followed, as nearly all major volcano eruptions have been, by a weakening of the monsoons upon which India and much of Africa depend for their crops. For these and several other parts of the world, the attempt to avert climate change by this route will bring on the very change it aims to prevent.[38]

Notes and references

[1]   Tony Blair, 'Breaking the Climate Deadlock: A Global Deal for Our Low Carbon Future' (The Climate Group, Exhibit 8), 23–24, available at https://www.theclimategroup.org/sites/default/files/archive/files/BTCDJune08Report.Fin.pdf (accessed on 10 July 2017).

[2]   David King, John Browne, Richard Layard, Gus O'Donnell, Martin Rees, Nicholas Stern and Adair Turner, 'A Global Apollo Programme to Combat Climate Change' (Executive Summary), 4, available at http://cep.lse.ac.uk/pubs/download/special/Global_Apollo_Programme_Report.pdf (accessed on 10 July 2017).

[3]   UN Climate Change: Climate Action, 'Lessons from Clean Development Mechanism Critical to Implementation of Paris Agreement' (13 May 2016), available at http://newsroom.unfccc.int/climate-action/lessons-clean-development-mechanism-implementation-paris-agreement/ (accessed on 10 July 2017).

[4]   Centre for Science and Environment, 'CDM: "Cheap Development Mechanism": Failed to Give Us the Transition to Clean Energy' (New Delhi, 2012, November), available at http://www.cseindia.org/userfiles/CDM_SAmedia.pdf (accessed 10 July 2017).

[5]   David King, John Browne, Richard Layard, Gus O'Donnell, Martin Rees, Nicholas Stern and Adair Turner, 'A Global Apollo Programme to Combat Climate Change'.

[6]   International Energy Statistics, 'Wind Electricity Net Generation 2014', available at https://www.eia.gov/beta/international/data/browser/#/?pa=00000000000000000000000000000001&c=ruvvvvv fvtvnvv1urvvvvfvvvvvvfvvvou20evvvvvvvvvvnvvuvo&ct=0&tl_id= 2-A&vs=INTL.37-12-AFG-BKWH.A&cy=2014&vo=0&v=H&sta rt=1983&end=2015 (accessed on 10 July 2017). Also see Tonto.eia. doe.govvia (accessed on 12 July 2017).

[7]   See http://sciencing.com/much-land-needed-windturbines-12304634. html (accessed on 10 September 2017).

[8]   Xi Lu, Michael B. McElroy and Juha Kiviluoma, 'Global Potential for Wind-generated Electricity', *PNAS* 27, (2009, 7 July): 10933–10938. The authors made this estimate on the basis of deploying 2.5 MW wind turbines over non-forested, ice-free, non-urban land areas of the world. They did not take offshore wind turbines into account.

[9]   Mark Z. Jacobson and Mark. A. Delucchi, 'Providing All Global Energy with Wind, Water and Solar Power, Part 1: Technologies,

Energy Resources, Quantities and Areas of Infrastructure, and Materials', *Energy Policy* 39 (2011): 1154–1169.

10   Carlos de Castro, Margarita Mediavilla, Luis Javier Miguel and Fernando Frechoso, 'Global Wind Power Potential: Physical and Technological Limits', *Energy Policy* 39 (2011): 6677–6682, available at http://www.eis.uva.es/energiasostenible/wp-content/uploads/2011/11/Global-wind-draft.pdf (accessed on 21 February 2014).

11   In actual fact, a wind's 'availability' of half the year has been recorded only in Denmark, Scotland and the north shore of Germany. In 2014, the global average was only 21.5 per cent.

12   Posted on 'The Oil Drum' on 5 September 2011, available at http://www.theoildrum.com/node/8322 (accessed on 8 May 2013).

13   *The Economist*, 'Sunny Uplands: Alternative Energy Will No Longer Be Alternative' (21 November 2012).

14   As a result, between 2009 and 2014, the cost of production of solar PV modules declined dramatically. In January 2015, Deutsche Bank predicted that by 2017 the total cost of a solar PV installation would fall to $1.77 per watt, but by then in India solar PV systems were being installed on rooftops for $1.20 per watt. The price was explicitly given in a press release issued by the Delhi state government on 12 September 2015. *Times of India*, 13 September 2015.

15   See https://www.bloomberg.com/news/articles/2016-01-19/solar-power-costs-drop-to-record-in-rajasthan-auction-in-india (accessed on 10 July 2017).

16   See https://www.bloomberg.com/news/articles/2016-12-15/world-energy-hits-a-turning-point-solar-that-s-cheaper-than-wind (accessed 10 July 2017). This is almost certainly a consequence of innovative financing that has brought down the cost of capital dramatically.

17   First Solar, 'First Solar Sets Thin-film Module Efficiency World Record of 17.0 Per cent. 17.0 Per cent Total Area Efficiency Module Confirmed by NREL', available at http://investor.firstsolar.com/releasedetail.cfm?ReleaseID=833971 (accessed on 31 January 2017).

18   Renewable UK: The Voice of Wind and Marine Energy. http://www.bwea.com/marine/resource.html

19   Alison Holm, Leslie Blodgett, Dan Jennejohn and Karl Gawell, 'Geothermal Energy: International Market Update' (Geothermal Energy Association, May 2010), 4. Also see Kirsten Korosec, 'Google

Project Maps US Geothermal Energy Potential', *Smart Planet* (25 October 2011), available at http://www.smartplanet.com/blog/intelligent-energy/google-project-maps-us-geothermal-energy-potential/9934 (accessed on 9 May 2011).

20   Alison Holm, Leslie Blodgett, Dan Jennejohn and Karl Gawell, 'Geothermal Energy: International Market Update', 7.

21   Massachusetts Institute of Technology, 'The Future of Geothermal Energy: Impact of EGS on the United States in the 21st Century', available at http://geothermal.inel.gov/publications/future_of_geothermal_energy.pdf (accessed on 10 July 2017).

22   Ibid.

23   Alison Holm, Leslie Blodgett, Dan Jennejohn and Karl Gawell, *Geothermal Energy: International Market Update*, May 2010.

24   Alison Holm, Leslie Blodgett, Dan Jennejohn and Karl Gawell, 'Geothermal Energy: International Market Update', 4.

25   See http://automobiles.honda.com/fcx-clarity/ (accessed on 10 July 2017).

26   John Voelcker, 'Honda Ends Three Green Models for 2015: Insight, Fit EV, FCX Clarity', Green Car Reports (29 July 2014), available at http://www.greencarreports.com/news/1092683_honda-ends-three-green-models-for-2015-insight-fit-ev-fcx-clarity (accessed 20 August 2014).

27   United States Environmental Protection Agency and US Department of Energy, 'Recently Tested Vehicles', fueleconomy.gov (23 December 2014), available at http://www.fueleconomy.gov/feg/fcv_sbs.shtml (accessed on 24 December 2014).

28   James Ayre, 'Toyota to Lose $100,000 on Every Hydrogen FCV Sold?' *CleanTechnica.com* (19 November 2014).

29   Guy Turner, 'European and Global CCS: The Current Status and Future Need', New Energy Finance (29 April 2009), available at http://ccs.newenergyfinance.com/ (accessed 10 July 2017).

30   This is described in detail in Chapter 8.

31   Gwynne Dyer, *Climate Wars—The Fight for Survival as the World Overheats* (Oxford: One World Press, 2011).

32   Fiona Harvey, 'Global Warming Crisis May Mean World Has to Suck Greenhouse Gases from Air'.

33   Tim Flannery, *A Third Way to Fight Climate Change*.

[34] Heather D. Willauer, Dennis R. Hardy, Kenneth R. Schultz and Frederick W. Williams, 'The Feasibility and Current Estimated Capital Costs of Producing Jet Fuel at Sea Using Carbon Dioxide and Hydrogen', and John Morgan, 'Zero Emission Synfuel from Seawater'.

[35] Tim Flannery, *Atmosphere of Hope: Searching for Solutions to the Climate Crisis* (New York, NY: Atlantic Monthly Press, 2016).

[36] Tim Flannery, *A Third Way to Fight Climate Change*.

[37] Long Cao and Ken Caldeira, 'Atmospheric Carbon Dioxide Removal: Long-term Consequences and Commitment', *Environmental Research Letters* (17 June 2010), available at http://iopscience.iop.org/article/10.1088/1748-9326/5/2/024011 (accessed on 10 July 2017). In their abstract, they write:

> In our extreme and idealized simulations, anthropogenic CO_2 emissions are halted and all anthropogenic CO_2 is removed from the atmosphere at year 2050 under the IPCC A2 CO_2 emission scenario when the model-simulated atmospheric CO_2 reaches 511 ppm and surface temperature reaches 1.8°C above the pre-industrial level. In our simulations a one-time removal of all anthropogenic CO_2 in the atmosphere reduces surface air temperature by 0.8°C within a few years, but 1°C surface warming above pre-industrial levels lasts for several centuries. In other words, a one-time removal of 100% excess CO_2 from the atmosphere offsets less than 50% of the warming experienced at the time of removal. To maintain atmospheric CO_2 and temperature at low levels, not only does anthropogenic CO_2 in the atmosphere need to be removed, but anthropogenic CO_2 stored in the ocean and land needs to be removed as well when it outgasses to the atmosphere. In our simulation to maintain atmospheric CO_2 concentrations at pre-industrial levels for centuries, an additional amount of CO_2 equal to the original CO_2 captured would need to be removed over the subsequent 80 years.

[38] Naomi Klein, *This Changes Everything*, op. cit.

Part II

What Can Save Us

Chapter Four

Solar Thermal Revolution

The only source of energy that can lift both the threats that hang over our heads is the sun. Fortunately solar energy is the area in which the most promising breakthroughs have occurred. But while solar photovoltaic power has stolen the limelight, the promise of a complete energy shift lies in concentrated solar thermal power, with solar PV playing a complementary role. Unfortunately, an attack on solar thermal power by advocates of solar PV is threatening to snuff out this possibility too. This assault reveals the worst features of the market economy.

The breakthrough that can replace both coal and nuclear power has taken place not in SPV but in solar thermal power generation. It has happened there because while scientists have yet to find an economical way of storing electricity for release when the sun is not shining, solar thermal power has crossed this barrier by storing the sun's energy as heat. The capacity of concentrated salt solutions to store heat up to 130°C had been known for a long time. But the key breakthrough came in the late 1980s when scientists found a way to store the heat of the sun for long periods without much loss of heat. Experimenting with different ways of storing heat at higher temperatures, scientists found that molten salt could store heat at temperatures up to 800°C for long periods.[1] This made it possible for concentrated solar power (CSP) plants to supply electricity when the sun did not shine.

As their name implies, concentrated solar thermal power plants use the simplest, and oldest, of stratagems to generate electricity. By focusing sun's heat, gathered over a large receiving area, onto a single receiver, they generate steam needed to drive turbo generator sets and supply electricity. The first experiments in concentrating the sun's rays to produce steam for power generation date back to the 1860s, and the first actual application of solar power was to run a 6 HP pump in 1904. Seven years later the journal *Popular Mechanics* reported that a 25 HP solar-powered irrigation pump had been developed that had pumped up 3 gallons of water per hour for several weeks during trial demonstrations in the United States, before being shipped to Al Maadi, in Egypt, to provide irrigation.[2] But large-scale solar power generation began in earnest in the Mojave Desert in California in the 1980s, spurred by the sharp rise in oil price in the 1970s.

The first commercial use of CSP technology to generate electricity dates back to the 1980s, when an American company BrightSource Solar Electric Generating System (SEGS) built, or took over, a cluster of nine plants in the Mojave desert in the United States, with an installed generating capacity of 354 MW. These plants did not have any means of storing heat so, like SPV plants, these were able to run on direct sunlight for only a little over five hours a day, or 1,840 hours a year. To meet a part of the peak demand it had contracted to meet, and reduce its overall generation cost, SEGS ran (and continues to run) its turbines on natural gas during the night.

By the end of the decade, heavily subsidised solar thermal power plants were delivering power to the grid in California, Israel and Australia. In 2008, there were six operational solar thermal plants in the world of a commercial size (against three SPV plants) supplying power to the grid. Another 16 were under construction, and 62 were being planned, mainly in the United States and Spain.

After the breakthrough in heat storage, CSP plants have gone through a constant process of evolution. The oldest commercial-scale plants use long, mirrored, parabolic troughs with a tube running through the centre which carries the fluid to be heated. But two more recent designs offer much higher levels of efficiency. These are the central power tower system and the parabolic dish system. The former focuses concentrated solar energy on a tower-mounted heat exchanger, also called a receiver. This system uses thousands of flat, sun-tracking mirrors called heliostats to reflect and concentrate the sun's energy onto a central receiver at the top of a central tower. The latter puts the receiver at the focal point of a parabolic dish and collects its energy from an array of such dishes. All three use computerised controls to make the mirrors face the sun throughout the day. While the parabolic trough system can heat the heat transfer fluid only to 400°C, the central tower system can concentrate the sun's energy as much as 1,500 times and achieve temperatures in the receiver of more than 1,000°C. Parabolic dish solar power systems are equally efficient but cannot, individually, generate more than 50 to 60 KW of power, so large numbers have to be aggregated to create commercial-sized power plants.[3]

CSP's advantages over SPV

CSP plants have many intrinsic advantages over their PV counterparts that lower their cost to society, if not to the investor. The first is that they need no REs such as tellurium and cadmium, which are toxic and whose supply is severely limited—tellurium being three times rarer than gold.[4] Second, solar concentrators are made of simple materials such as mirrors, mild and stainless steel and do not require the sophisticated, energy-intensive manufacturing processes that PV panels require. Third, unlike SPV, the performance of CSP plants improves when the ambient temperature rises. This makes them especially suited for installation in tropical and equatorial, desert and arid regions, where most of the world's poor reside. Fourth, as mentioned above, CSP power needs no special steam or gas turbines. The same turbines that generate coal and nuclear power do so with equal ease from solar thermal power. So feeding this power into the grid does not require stepping up its voltage as PV power needs. Therefore, replacing coal or nuclear power with solar thermal collectors does not raise the cost of transmission.

But CSP's most important advantage—the one that can, and almost certainly will, change the world's future—is that it can store the sun's energy at negligible cost not as electricity but as heat. In the late 1980s, scientists found that a combination of molten sodium and potassium nitrates could retain 95 per cent or more of the heat fed into them for 24 hours or longer. This is what has made CSP plants capable of supplying power on demand and therefore replacing not only coal but also nuclear and large hydro-electric power plants.

Phase I—Parabolic trough

The first commercial plant that relied on stored heat to meet peak consumer demand was Andasol 1 in Spain. Andasol 1 had a generating capacity of 50 MW and sufficient extra heliostats for collecting heat to run the turbines for 7.5 hours in the absence of direct sunlight. This raised the expected plant availability to 3,160 hours a year, that is, 36 per cent of the hours in a year, almost double of what SPV is capable of giving. But at that early stage of development, Andasol's cost turned out to be very high.[5] Some idea of the subsidy it needed may be had from the fact that the Spanish government agreed to buy power from it at 27 centimes (34 US cents) per KWh.[6]

But since then the cost of generation has fallen dramatically. Andasol 1 was only the first of three identical plants. Since 2008, the investors, the ACS/Cobra group, have built two more sister plants, Andasol 2 and 3, at the same location and endowed them with up to 12 hours of energy storage capacity. The additional storage capacity brought the three plants' availability up from 36 to 41 per cent of the year, that is, 3,560 hours in a year. Andasol, therefore, became the first commercial plant that was capable of guaranteeing the supply of electricity 'during all peak hours in most weather conditions'. It, therefore, went a long way towards finding the holy grail of energy developers and being able to guarantee power on demand.

The success of Andasol 1 triggered a sharp revival of interest in solar thermal power plants. And the new plants are getting steadily bigger. In 2010, the Spanish renewable energy pioneer, Abengoa-Solana began to build a 280 MW plant also using Andasol's parabolic trough technology not far from Phoenix, Arizona. The plant, named 'Solana', is designed for generating peak-load power for 12 hours a day, of which 6 hours is from stored heat[7] and is selling all of it to the Arizona Public Service, the state's largest public utility, to meet its peak power requirements. The levelised cost of electricity from this plant has been estimated to be 19.2 cents per unit.[8] In 2013, this was 'less' than the cost of peak power generated from natural gas.[9] But it was still far higher than the cost of base-load power from coal or gas.

Phase 2—Central tower power

But Andasol and Solana showed that the parabolic trough method of power generation would only be profitable at peak-load tariffs. Therefore, they had not broken the final barrier to the replacement of coal. This was done by Terresol, a Spanish company that has been one of the pioneers in solar thermal power, when it commissioned a 20 MW central tower plant, named Gemasolar, outside the town of Fuentes de Andalucía, in Spain. Gemasolar's construction was approved in February 2009 and it began trial production in May 2011.

Gemasolar is the first solar plant in the world that is capable of providing power on demand, day and night, in exactly the same way as coal-fired plants do today. To do this, it stores enough of the sun's energy to generate 15 hours of power without the aid of direct sunlight.[10] It does this by maintaining a molten mixture of the two salts, sodium and potassium nitrate at around 290°C, which is pumped to the top of a central tower where it is bombarded with the sun's heat by concentric rings of large, flat, mirrors called heliostats that follow the sun through the day with the aid of computerised controls. Systems using molten salt are, at present, pushing the temperature of the salt to 565°C or a little more, but they can push it up as high as 1,000°C, which is far beyond the 590°C needed to generate supercritical steam in the most advanced boilers currently in use in coal-based and nuclear plants.

This has enabled Gemasolar to generate up to 5,475 hours of power a year from stored heat alone. To meet its target of providing 6,500 hours of guaranteed power to Fuentes de Andalucia, therefore, it needs to generate power directly from the sun for no more than 3 hours a day.

Torresol celebrated Gemasolar's second anniversary in October 2013 by providing 24 hour-power to Fuentes de Andalucia for 37 days non-stop. It is therefore the first solar power plant that has been able to completely replace fossil fuels in the generation of electricity for any length of time.[11] Two years after that event, Gemasolar's capacity to supply power on demand during the summer, when days are long, is no longer in doubt, but it still needs some fossil fuel backup in winter. One way around this problem would be to build hybrid solar thermal and PV plants in which the much cheaper-to-install PV plants provide the extra few hours of power.

Not surprisingly, the next few CSP plants have chosen the central tower system. In the United States, the first large commercial plant using this system is the Crescent Dunes plant at Tonopah, in Nevada. With a generation capacity of a 110 MW, it was built in two years and generated power on a trial basis in December 2013. Crescent Dunes is equipped with 10 hours of heat storage capacity and will deliver electricity for 14 hours a day between noon and 2:00 AM at night.[12] Under a 25-year agreement, its promoters agreed to deliver power to Las Vegas at 13.5 cents per KWh (but a similar plant being set up in South Africa by the same company is expected to produce power at 12.4 cents a unit).[13]

Crescent Dunes has been followed by a 292 MW CSP plant built around three central power towers, at Ivanpah, in the

Californian part of the Mojave Desert, 40 miles from Las Vegas. Ivanpah is so far the largest completed CSP plant built anywhere in the world.[14] Although begun later than Crescent Dunes, it became, in February 2014, the first American CSP facility to start supplying power to the grid. But teething troubles and environmental clearances have held up full-scale commercial production at both plants. Crescent dunes had not begun supplying power to the grid even in May 2015.

Teething troubles

It needs only one successful commercial plant to prove the viability of a technology and eliminate most of the risk of investment. The Gemasolar plant has done that for solar thermal power. But there has also never been a new technology that has not encountered teething troubles, and stored heat CSP plants in the United States have proved no exception. Unlike Gemasolar, the success of which is built upon more than a decade of learning from older generation solar thermal plants in Spain, Crescent Dunes and Ivanpah were started within months of each other. So both were pioneer ventures that had to do all the learning that Terresol, the builders of Gemasolar, and Abengoa, the builders of Solana, had done over more than a decade in Spain. Teething troubles were, therefore, only to be expected. In the United States, Ivanpah, the first large CSP plant to go on stream, using the central tower system of heat collection, has had the most, and therefore borne the brunt of the criticism that has followed. By March 2015, it had achieved only 40 per cent of its rated capacity.[15]

The management blamed adverse weather conditions and jet contrails for lowering its efficiency but, apart from normal start-up problems such as leaky heat receiving tubes and malfunctioning steam turbines, Ivanpah owed most of its poor initial performance to poor design. The project's engineers, BrightSource Energy, relied upon a technology called the Luz Power Tower 550 that they were familiar with. So the fluid they heated in the central tower receiver was not molten salt but water. As a result, the plant has no heat storage capacity and had to reheat the water every morning to get a supercritical head of steam. BrightSource's engineers underestimated the amount of gas they would need to do this by a factor of four! Having underestimated the gas they needed they inevitably underestimated the time it would take to acquire the critical head of steam.

The other main design flaw was an assumption that the plant would draw power 31.4 per cent of the year—close to 8 hours a day—from the sun. This was over-optimistic, to say the least, because at a similar latitude in Spain, Andasol 1 had been able to deliver power for 35 per cent of the year only with the help of six hours of stored heat.[16] The Mojave Desert gets a lot more sun than southern Spain, but even allowing for the difference, the assumption was unrealistic.

A motivated backlash—Drawbacks of an unguided market

Teething troubles are normal, and design mistakes get remedied in the next plants to be put up. They, therefore, need to be treated with a degree of indulgence. But solar thermal power, in general, and Ivanpah, in particular, have been denied this indulgence. Instead, from the day these CSP plants began to be constructed they have been the targets of an unrelenting attack by the votaries of SPV power. Many of the concerns raised in the media about Ivanpah were legitimate. Birdwatchers voiced unease over the death of large numbers of birds that have strayed into the intense solar radiation field of these plants. Nature lovers protested that it would destroy the habitat of the Arizona Desert tortoise. The greatest cause of unease was that solar thermal power plants also consumed millions of gallons of water a day to cool the mirrors and keep them free of dust.[17]

The investors responded to these with changes of design: to save the habitat of the Arizona Desert tortoise the size of the plant was reduced from 440 MW, with four power towers to 297 MW with three. When 150 birds were killed during a pre-commissioning test at the Crescent Dunes plant in January 2014, the authorities changed the configuration of the mirrors for future tests to ensure that the solar radiation generated was widely dispersed.[18] This ended bird kills caused by the testing process and greatly reduced the loss of birdlife at other times.[19] To save water, the engineers replaced wet cooling by dry cooling. This has reduced the power output by 2 to 3 per cent of the energy generated but reduced the water requirement by 95 per cent to less, the engineers claim, than what is consumed by a neighbouring golf course.

In spite of this, Ivanpah's specific problems were graduated by its critics into generic defects of all solar thermal plants. These are that CSP plants are (a) dangerous to avian and other forms of wildlife; (b) they are more complicated (than SPV) and therefore more prone to disruption;[20] (c) they consume vast amounts of water; (d) they only work in direct sunlight because even thin clouds and contrails can bring their power output down sharply. The most important criticism, however, was that they are far too expensive compared to SPV power.

What the barrage of criticism has obscured is that Ivanpah is sensitive to changes in solar radiation only because it has been designed to heat water and not molten salt in the receiver at the top of the central tower. In plants that use molten salt, and those which are equipped with storage tanks, a dip in the temperature of the molten salt exposed to direct sunlight can be instantly corrected by releasing some of the molten salt in the storage tank into the heat exchange boilers. What is more, because of its very high thermal capacity the temperature of molten salt in the receiver does not fluctuate with changes in solar radiation the way that the temperature of steam fluctuates when water is the heat transfer fluid.

That this attack was being orchestrated became apparent when a vitriolic article appeared on an online website 'Wired', which dwelt with glee upon a fire that broke out on one of Ivanpah's three central towers in May 2016 because of a misalignment of the heliostats. Not only did it describe the concentrated heat beam generated by the solar field a 'Death Ray' but it made no secret of its strong advocacy of SPV power. Buried in the article was one significant paragraph:

> Ivanpah's biggest problem, though, is hard economics. When the plant was just a proposal in 2007, the cost of electricity made using Ivanpah's concentrated solar power was roughly the same as that from photovoltaic solar panels. Since then, the cost of electricity from photovoltaic solar panels has plummeted to 6 cents per kilowatt-hour (compared to 15 to 20 cents for concentrated solar power) as materials have gotten cheaper. 'You're not going to see the same thing with concentrated solar power plants because it's mostly just a big steel and glass project', says Schultz. It can only get so much cheaper.[21]

Present in this remark is an invitation to believe that there is literally no limit to how much further the cost of SPV panels can fall; and absent from it is any acknowledgement that more than half the cost of SPV systems is incurred in their installation. Since these cannot be brought down easily, there is a floor below which the cost of SPV power too cannot sink.

But what is most telling is the complete absence of even a hint that, no matter how cheap it becomes, and no matter how smart power grids become, SPV power will *never* be able to meet the demand for base-load power. For there is not enough of tellurium in the world for the PV panels, and Lithium for batteries, to generate the sheer volume of electricity being generated by coal. The most efficient PV panels in use in 2016 required 93 tonnes of tellurium per GW of generating capacity. The total consumption of electricity in that year was of the order of 22,000 TWh, of which three quarters was generated from fossil fuels. To replace this with SPV power will require a minimum of 8 TW of solar-generating capacity. Since the most efficient panels being used today require 93 tonnes of tellurium per GW of generating capacity, to replace fossil fuels entirely will require 750,000 tonnes of tellurium or thereabouts. Based on the current production of copper (of which tellurium is a by-product) its availability is about 800 tonnes a year (although no more than 135 tonnes is actually being produced). So even if all the tellurium in the world were to be reserved for manufacturing SPV panels, their efficiency will therefore have to be increased more than 200 times to meet even the current demand for fossil fuel-based power. But since solar panels are already converting 17 per cent of the energy they receive from the sun, the maximum this can be increased by, even in theory, is no more than five times.

The second assumption upon which the advocacy of SPV power is based—that with battery costs plummeting, SPV power with battery storage will soon be cheaper than conventional power—is also wishful thinking, for a report released by the World Energy Council in January 2016 estimated that the absolute minimum that storage would cost in 2030 is likely to be 5 cents per KWh.[22] Add this to the 3 cents that PV power is being priced at today and allow for a further halving of solar thermal power costs in the next 13 years, and the price advantage over stored solar thermal energy will disappear completely.

In fact, the mere effort to expand SPV and battery storage beyond a point will reverse the relationship between the two forms of power, for there simply is not enough lithium on the planet to meet the demand for batteries to fuel 1.2 billion electric cars and for providing 8 TW of stored power for 16 hours a day every day of the year. The US Geological Survey has estimated that the global reserve of lithium is of the order of 13.5 million tonnes. If the number of automobiles is frozen at today's 1.2 billion and each electric car battery uses up to 10 kg of lithium, as the Toyota Tesla does, the auto industry alone will require 12 million tonnes of lithium without allowing for annual replacement and fleet growth. So without some extraordinary leap in technology, of which there is no sign whatever so far, electric cars are yet another huge blind alley down which the giant global corporates are dragging the world today.[23]

Notwithstanding these hard realities, the barrage against solar thermal power has achieved its purpose, for most fund managers are convinced that solar thermal power will never be cheaper than PV and might not even be cheaper than coal-based power except when generated in desertified areas of the world where solar radiation is very high and the sky remains clear for all but a few days in the year. They admit that SPV cannot by itself meet the demand for base-load power, but believe that with the dramatic fall in the price of batteries that is now taking place, the future will lie in a combination of PV generation and battery storage.

A struggle that can cost humanity its future

The developing clash between solar thermal and SPV power generation is uncannily similar to the clash between ethanol and methanol as alternates to gasoline that is described in the next chapter. The two controversies reveal one of the worst traits of the market economy, which is to force the belief relentlessly down peoples' throats that the market always knows best, because it fosters competition and competition will ensure the best possible outcome for society, even where it is patently obvious that it will do the opposite. In this case, one needs to ask why the advocates of SPV power insist on using the price of coal, and not oil or gas-based power as the yardstick against which to assess economic viability when, in energy-equivalent terms, oil is seven times more expensive than coal.[24] The answer would probably be that most power generation is based on coal, and that the prices of all three are set by demand and supply. But the wide disparity suggests the presence of implicit subsidies. The least of these is the cost of maintaining the townships, and providing health and education in coal townships, all of which is borne by the community out of wages and taxes and does not therefore figure in the coal mining companies' private costs.

If the slightest adjustment is made to take these into account, then both SPV and solar thermal become comfortably viable. That, and not the need to 'punish carbon emitters', is the rationale for the carbon tax that the Paris summit failed to adopt, but which was proposed, in a revolutionary volte-face in the United States by the Climate Leadership Council, a group of republicans headed by former secretary of state James Baker, for consideration by the Trump administration.[25]

The world needs both SPV and CSP power. So urgent is humanity's need to find substitutes for coal and natural gas to generate both base-load and peak power that there is abundant room for both technologies to develop. Indeed combining both in hybrid solar power plants promises to bring capital cost down sharply and do away with the need for a fossil fuel backup altogether, at the same time. But instead of promoting such a synergy, the vast majority of published articles are pitting the two technologies against each other and vociferously demanding that readers choose one over the other.

Cost of generation—Comparison against what?

In concentrated solar power generation 'Power Tower' technology has opened up a virtually limitless opportunity to increase output and lower costs in the future. While the maximum concentration of solar thermal energy possible with parabolic troughs is 212 times, with heliostats it is 44,000 times.[26] This is because every 1 per cent increase in the sharpness of focus increases the energy delivered to the receiver by 10 times as much.[27]

Nearly all of the decline in the cost of CSP power over the past decade has come from the improved concentration of solar radiation on the receiver. The highest that existing parabolic trough systems are able to raise the temperature of the steam fed into the turbines is about 400°C. But central tower systems are operating at 565°C. This temperature is sufficient for the supercritical steam turbines which are now being used by advanced coal-fired power plants. Further increases in temperature, till 1,000°C, are possible but unnecessary because the ultra-supercritical boilers and turbines now coming into use require steam at a temperature of only 593°C. These boilers and turbines give about one-sixth more power than the best sub-critical boilers and turbines in use today.[28]

A host of other design improvements in the central tower heat collection system, such as taller central towers, the replacement of single pole supports for the heliostats with carousel mountings, and the shifting of the tracking motor from the heliostat housing to the carousel can reduce the cost of the support structure by a third; while simultaneously sharpening the focus of the heliostats. They do this by replacing special with mild steels and reducing the amount of steel and cement needed by a third, by reducing the vibration of the heliostats caused by wind and, above all, by eliminating the vibration caused by the tracking motors which are completely separated from the heliostat.[29] This is permitting an exponential increase in the energy delivered to the receiver, and a corresponding reduction in the heliostat surface area for any given power output.

In a detailed presentation made to the National Renewable Energy Laboratory in Boulder, Colorado, a Spanish firm Titan Tracker showed that its 'enhanced carousel' design was four times more accurate than the single pole mounted heliostats used by the first generation of CSP plants.[30]

In developing countries, much lower costs of construction will bring the cost of electricity down further. A back-of-the-envelope cost calculation shows that a clone of the Gemasolar plant, set up in the Thar Desert of Rajasthan without any of the improvements suggested above, would be able to supply power at around 9 cents per KWh.[31]

In spite of global recession, the collapse of raw material prices after China's fiscal stimulus programme ended, and the raging conflict in the middle east, solar thermal power generation is spreading to other countries. In July 2013, there were solar thermal power plants with a total capacity of 1,300 MW in operation, an additional 2,300 MW was under construction and 31,700 MW in an advanced planning stage. Since then many more large CSP plants have begun to generate power. The economics of CSP power generation is therefore much better understood.[32]

Economic viability

The Gemasolar plant is estimated to have cost 235 million Euros cost, and is built on 195 hectares of land. Although its levellised (equated over plant life) cost of generation is not generally known, it is possible to make an approximate estimate with the help of the guidelines laid down by the US National Renewable Energy Laboratory. This comes to about 19 centimes (currently US 22 cents) per unit. Like other solar power plants in Spain, therefore, Gemasolar has needed a substantial subsidy from the Spanish government to be economically viable.

Europe is well known, however, for its high construction costs. A recent, detailed, American study of the cost of building nuclear power plants in Europe, the United States and China has shown that these are one and a half times as expensive to build in Europe than in the United States and three times as expensive as identical reactors built in China. A similar cost gap exists in solar panels and heliostats. With the entry of the Chinese into this market, the cost of solar power generation equipment has fallen by 60 per cent. In 2014, Chinese companies were advertising the availability of heliostats for CSP plants at $120/m². As a result, the cost of solar thermal power too is falling rapidly in many parts of the developing world.

A back-of-the-envelope calculation for a Gemasolar clone set up in India, based on the assumptions that heliostats cost $120/m² and account for 40 per cent of the cost of a CSP plant; employ 90 persons on a permanent basis (against Gemasolar's 45), at an average salary of $6,000 per year (about a fifth of the

European average), suggests that even after allowing for higher interest rates (7.5 per cent) and a plant life of 30 years, the levellised cost of electricity supplied to the grid in India will be at most 8 to 9 cents per KWh.[33] This is fully competitive with modern coal-based power plants.

These developments have not gone unnoticed. Speaking at the *EuroScience Open Forum* in Barcelona, in 2008 Arnulf Jäger-Waldau of the European Commission's Institute for Energy said it would require the capture of just 0.3 per cent of the light falling on the Sahara and Middle East deserts to meet all of Europe's energy needs.[34] In 2009, the Desertec Foundation, created by Gerhard Knies, a German scientist, gave shape to a plan that he had been mulling over since the Chernobyl disaster of 1986. This was to replace nuclear power in Europe with solar power drawn from the Sahara and transmit it to Europe through a high tension grid. This plan gained momentum from Germany's decision to close down its nuclear power plants after the Fukushima disaster. By 2011, Desertec had created Dii Desertenergy, a consortium of companies that drew up plans to construct solar thermal power plants on land leased in Libya, Tunisia, Morocco, Egypt, Saudi Arabia and Syria. These plans were put on hold because of the turmoil that began with the onset of the Arab Spring and finally abandoned when global commodity prices, including those of the heavy metals used in PV panels and batteries, crashed in 2014. Only the Moroccan component of the original plan has survived.

Ouarzazate—The path to the future

Ouarzazate, in Morocco, is the place where the world's largest solar thermal power plant is nearing completion. With a planned generating capacity of 580 MW, it is being built in three phases with a conservative respect for technology, and a disregard for initial costs that should be an object lesson for those who want to save the world, but only if it is the market economy that does the saving.

The first phase, Noor 1, which has a generation capacity of 160 MW was switched on by King Mohammed VI on 5 February 2016.[35] It has three hours of storage, costs $840 million, took 2 years and 8 months to build from groundbreaking to commissioning and is selling its power to the grid under a 25-year power purchase agreement producing power at 18.9 cents a unit. In this phase, the designers have relied upon the older, tried and tested parabolic trough design pioneered by Andasol but, learning from Ivanpah's mistake and Gemasolar's success, have used molten salt as the heat transfer medium instead of water. The second phase, Noor 2, which also uses the parabolic trough design, has a generating capacity of 200 MW, provides for 7.5 hours of heat storage and is costing $1,041 million to build. Consequently, its cost of generation has fallen to 15.6 cents a unit.

It is only in Noor 3, which has a capacity of 150 MW, that the central tower collection system is being used. At 216 m, its central tower receiver will be the highest in the world, and each of its heliostats is the size of half a tennis court. Noor 3 will also be selling its power at 16 cents a unit, but the main investor ACWA Power is building a similar plant. The three phases are designed to store heat for 3, 7.5 and 8 hours, respectively. There is also an 80 MW SPV plant that is designed to supplement power supply during the day time and extend the storage capacity of the main plant.

In a market economy, no one would have touched the project with a bargepole. It has only come into being only because of the personal backing of the King, that is, the state of Morocco, to end its utter dependence upon imports to meet its energy needs. Morocco depends on imports for 97 per cent of its energy needs. In spite of a 7 per cent annual increase in its energy consumption, the Ouarzazate power station will reduce import dependence to 58 per cent by 2020. Ouarzazate shows the world that the market cannot save mankind. Only concerted action by the state, using the instrumentality of the market, can do so.

ACWA Power is building two more 300 MW solar thermal plants in Saudi Arabia and Dubai and has quoted a price of 12 cents a unit for the former.

Solar power can replace hydro-power right away

Solar thermal power may still be some distance away from under-cutting heavily subsidised coal-based power. But why is no one comparing its cost to the cost of hydroelectric power? For by any reckoning it is a good deal cheaper already. As estimates given below show, not only is the bare investment per delivered unit of power much lower, but doing so will avoid the immense cost that society has to pay whenever the earth is ripped apart to build a dam with its attendant tunnels, power stations and roads.

Nor is this the only type of social cost that hydro-power can exact. Another cost that is looming ever larger as the world begins to reach the limits of its finite resources is that of war between riparian states over control of the flow of the river waters. Four large countries are blessed with enormous renewable reserves of water. These are Brazil, Russia, the United States and Canada. A fifth, China, has ample water for its needs, but so unevenly dis-tributed that it too is taking decisions that are bringing it into conflict with its neighbours. A far larger number of countries are chronically short of water and are already suffering 'a perfect storm of rapid population growth, resource depletion, poor governance, economic stagnation, and unsettling climate change'. The most fragile among them are concentrated in a vast swathe stretching from North Africa across the Middle East and Horn of Africa into Central, and South Asia, precisely the area which is being torn apart already by conflicts over religion and oil.[36]

There are, at present, 37 ongoing disputes over water, and many more are brewing. As the scarcity of water increases, whether through increased consumption or accelerated desertification, these will become more intense, and new ones will spring up.

Solar power cannot alleviate absolute shortages of water, but it can greatly reduce the conflicts that will arise over control of its flow between countries. For most of these do not arise from a shortage of drinking water, but attempts to generate power from it.

Environmentalists, seismologists and human rights activists in India, for instance, have been questioning the utility of hydro-electric power projects for several decades but so far the engineering and heavy industrial lobbies that benefit from investments in infrastructure have always won the day by claiming that hydro-power is an environment-friendly alternative to thermal power, that it is needed to meet peak loads and thereby maximise the efficiency of thermal power plants and that although their capital cost is high, their operating cost, when compared with coal-fired plants, is very low.[37] China has justified the renewed emphasis on hydro-power projects in its 12th Energy Plan on almost the same grounds.

These claims are more propaganda than fact for most such projects have cost more to build and have delivered less power than their promoters had promised. In India, for example, a standing committee of the Parliament found, in 2007, that the hydel schemes implemented in the previous 60 years had cost 3.88 to 25.16 times more than the original project estimates and had taken nine to sixteen years longer to complete than the government had anticipated.[38]

Engineers have also propped up the myth that hydro-power is cheaper in the long run than coal-based power by inflating the number of hours in the year for which such plants can deliver electricity. The normal assumption made in appraisals of hydro-power projects in the Himalayas, for example, is that they can operate at 60 per cent of their capacity, that is, for 5,256 hours a year. No Indian, or for that matter Chinese, hydro-power plant has ever achieved this level of availability. In 2009–2010, Indian hydel plants generated power for an average of only 2,893 hours in the year, giving a plant availability of only 32 per cent. This has doubled the delivered cost of hydro-power.

Hydro-power does save CO_2 emissions, but only over a long period of time. Its immediate impact is to sharply increase CO_2 emissions. In a 10-year plan for developing hydro-power in the Himalayas, released in 2007, the Indian Central Electricity Authority had estimated that to add 15,000 MW of hydro-power (slated for the 11th Five Year Plan), it would need 30 million tonnes of cement and 15 million tonnes of construction steel.[39]

Time—The true cost of hydro-power

The true cost of hydro-power, however, is not money but time. Time does not have an economic cost so long as there are no alternatives available. But the demand for peak-load power, which is what hydroelectricity is usually intended to meet, can now also be met by wind, SPV and solar thermal power or a combination of the three. So time does have a value. Once this is factored in, there is no longer even a shred of economic justification left for hydel power. For while the minimum time needed to implement large hydroelectric projects is around 10 years, wind and SPV plants take less than two and solar thermal less than three years to build. So the shortage of peak-load power that hydel is intended to alleviate, and the production, savings and reinvestment that it is intended to facilitate, are available eight years sooner than hydel can provide them. Once the benefits foregone during these eight years are taken into account, every vestigial remnant of the case for hydel power flies out of the window. So does the case for nuclear and even coal-based power.

Comparing true costs

Inter-country comparisons of the cost of hydro-, solar power and nuclear-, and coal-based power generation are not easy to make because they are affected by country-specific factors such as differing rates of interest, differing costs of acquiring land, differing rates of inflation and fluctuations in the exchange rate. The estimates

Table 4.1: Comparison of true economic cost of alternative power sources, 2013

Components	Thermal	Nuclear	Hydro	Solar (CSP)[a]
Cap cost/MW capacity (US$m)	3.246	5.530	2.936	7.314[c]
Plant-load Factor (hr/yr)	6,400[b]	7,884[b]	2,893[b]	6.500[d]
Cap cost per MWh ($)	507	701	1,014	664
Construction period (yrs)	5	8	10	2.5[e]
Net saving foregone[f]	1,350	2,970	4,050	0
True Cap cost per MWh ($)[g]	*1,857*	*3,601*	*5,064*	*664*

Source: 'Overnight' or 'bare' capital costs obtained from US Energy Information Administration, 'Updated Capital Cost Estimates for Utility Scale Electricity Generating Plants'. These have been adjusted to obtain estimates for India (and China) in the ways described in the table's footnotes.

Notes: [a]Capital cost based upon the technical specifications of the Gemasolar 19.9 MW central tower CSP set up at Fuentes de Andalucía in Spain, which came into operation in May 2011.

[b]These PLFs are the actual experience in India. Nuclear power PLF applies to plants that did not experience difficulties in obtaining uranium.

[c]Based upon price of heliostats prices quoted by Chinese suppliers ($120/m²) and the assumption that these account for a little under half of the total cost of the solar thermal plant. In 2014, at least one American supplier is also offering these at $126. The cost of heliostats is falling rapidly and is expected to reach $70/m² in the near future.

[d]PLF based on Gemasolar's actual performance in 2013 and 2014. It has achieved this by supplying 15 hours of power per day from stored heat.

[e]Actual construction period at Fuentes de Andalucía was 2 years 3 months.

[f]This is calculated as the saving out of additional GDP that is foregone during the longer gestation period of the project. In India, the GDP in 2011–2012 was almost exactly $1.5 trillion. The saving rate was 36 per cent.

[g]This is row 4 + 6.

given in Table 4.1 are therefore based upon 'overnight', that is, bare capital costs, in one country, the United States, in April 2013.[40]

The table highlights the difference between the 'private' capital costs of different types of power plants, that is, the cost that appears in the company's accounts, and the true economic cost that is borne by the nation. With the cost of SPV panels having crashed and those of heliostats having fallen further in the last four years, the capital cost to the company of even solar thermal power is second only to that of a 'clean' coal-based thermal power plant and is marginally lower than nuclear power. The capital cost of wind or PV power is well below all three.

But a calculation of the capital cost to society, obtained by deducting the additional savings generated out of eight years of additional GDP, changes the balance between them dramatically. For it shows that the true economic cost of solar thermal power 'is already one-third of coal-based thermal power; one-sixth of nuclear power and one-eighth of hydro-power'. All of its other ecological and social benefits, described in the last part of this book, are a bonus.

Notes and references

[1] Nicole Pfleger, Thomas Bauer, Claudia Martin, Markus Eck and Antje Wörner, 'Thermal Energy Storage—Overview and Specific Insight into Nitrate Salts for Sensible and Latent Heat Storage', *Beilstein Journal of Nanotechnology* (13 February 2015), available at http://www.beilstein-journals.org/bjnano/single/articleFullText.htm?publicId=2190-4286-6-154 (accessed on 8 February 2017).

[2] 'Sun Power Operates Pumping Plant' in *Popular Mechanics* (December 1911), available at http://books.google.co.in/books?id=-t0DAAAAMBAJ&pg=PA843&dq=Popular+Mechanics+Science+installing+linoleum&hl=en&sa=X&ei=y4zsT9OSGMriqA-GU1PW8BQ&sqi=2&redir_esc=y#v=onepage&q&f=true (accessed on 10 July 2017).

[3] See http://www.solarthermalworld.org/sites/gstec/files/Solar%20Power%20India%20a%20Reality.pdf (accessed on 10 July 2017).

[4] Tellurium makes up a scant 0.0000001 per cent of the earth's crust, making it three times rarer than gold. Source: A Yale Environment 360 report (13 November 2013), available at http://e360.yale.edu/feature/a_scarcity_of_rare_metals_is_hindering_green_technologies/2711/ (accessed on 27 June 2017).

[5] NREL, 'Concentrating Solar Projects—Andasol 1'. See also, Clean energy case study: http://www.cleanenergyactionproject.com/CleanEnergyActionProject/Solar_CSP___Concentrating_Solar_Power_Case_Studies_files/Andasol%20Solar%20Power%20Station.pdf.

6   NREL, 'System Advisor Model (SAM) Case Study: Andasol 1' (Aldeire, Spain), available at https://sam.nrel.gov/sites/sam.nrel.gov/files/content/case_studies/sam_case_csp_physical_trough_andasol-1_2013-1-15.pdf (accessed on 10 July 2017).

7   *Clean Technica*, 'Green Jobs and Solar Power at Night. Abengoa's 280 MW Solana CSP Plant Nearly Ready', available at http://cleantechnica.com/2012/12/01/green-jobs-solar-power-at-night-abengoas-280-mw-solana-csp-plant-nearly-ready/ (accessed on 10 July 2017).

8   Michael Stavy, 'Levellised Cost of Electricity from the Abengoa's Arizona CSP Plant with Storage', available at http://www.nvo.com/stavy/nss-folder/ewec2010paper/Stavy.AEA.WSJ.12.31.10.Article.Solana.pdf (accessed on 8 February 2017). See also Rebecca Smith, 'Solar Plant to Generate Power After Sundown', *The Wall Street Journal* (31 December 2010), available at https://www.wsj.com/articles/SB10001424052970204204004576049571490850918 (accessed 10 July 2017).

9   Heinz-Muller Steinhagen, 'Concentrating Solar Thermal Power', *Philosophical Transactions of the Royal Society A*, Figure 6. The levellised cost of peak power in 2012 was 22 cents.

10   Torresol Energy, 'Gemasolar', available at http://www.torresolenergy.com/TORRESOL/gemasolar-plant/en (accessed on 10 July 2017). See also http://en.wikipedia.org/wiki/Gemasolar_Thermosolar_Plant (accessed on 10 July 2017). Gemasolar employs 2,650 heliostats with a surface area of 304,000 m² to do so.

11   Here are the technical specifications and estimated capital cost of the Gemasolar plant as provided by Torresol. Gemasolar is the first high-temperature solar receiver with molten salt, which provides 15 hours of thermal storage and an annual capacity factor of about 75 per cent.

Status Date:	24 October 2011
Technology:	Power tower
Status:	Operational
Country:	Spain
City:	Fuentes de Andalucía
Region:	Andalucía (Sevilla)
Lat/Long Location:	37°33' 44.95″ North, 5°19' 49.39″ West

Land Area:	195 hectares
Solar Resource:	2,172 KWh/m2/yr
Source of Solar Resource:	Sener
Electricity Generation:	110,000 MWh/yr (Expected/Planned)
Break Ground:	February 2009
Start Production:	April 2011
Cost (approx.):	230,000,000 Euro
Construction Job Years:	800
Annual O&M Jobs:	45

Plant Configuration
Solar Field

Heliostat Solar-Field Aperture Area:	304,750 m²
# of Heliostats:	2,650
Heliostat Aperture Area:	120.0 m²
Heliostat Manufacturer:	Sener
Heliostat Description:	Sheet metal stamped facet
Heliostat Drive Manufacturer:	Sener
Tower Height:	140 m
Receiver Manufacturer:	Sener
Heat Transfer Fluid Type:	Molten salts (sodium and potassium nitrates)
Receiver Inlet Temp:	290°C
Receiver Outlet Temp:	565°C
Receiver Temp. Difference:	275°C

Power Block

Turbine Capacity (Gross):	19.9 MW
Turbine Capacity (Net):	19.9 MW
Cooling Method:	Wet cooling
Fossil Backup Type:	Natural gas
Backup Percentage:	15%

Thermal Storage

Storage Type:	2-tank direct
Storage Capacity:	15 hour(s)
Thermal Storage Description:	One cold salts tank (290°C) from where salts are pumped to the tower receiver and heated up to 565°C, to be stored in one hot salts tank (565°C). Annual equivalent hours = 6,500.

[12] It will generate 482,000 MWh of power a year and achieve a plant-load factor (PLF) of 52 per cent. Noon to 2:00 AM are the peak-load hours at Las Vegas.

[13] Giles Parkinson, 'World's Biggest Solar Tower + Storage Plant to Begin Generation This Month', *REneweconomy* (14 May 2015), available at http://reneweconomy.com.au/2015/worlds-biggest-solar-tower-storage-plant-to-begin-generation-this-month-22860 (accessed on 10 July 2017).

[14] The investors, BrightSource and NRG Energy, have set up three power towers of approximately the size of Crescent Dunes, to generate 292 MW of electricity.

[15] Cassandra Sweet, 'High-tech Solar Power Projects Fail to Deliver', *The Wall Street Journal* (12 June 2015), available at http://www.wsj.com/articles/high-tech-solar-projects-fail-to-deliver-1434138485 (accessed on 10 July 2017).

[16] 1.5 hours were kept in reserve for emergencies.

[17] Basin and Range Watch, 'Crescent Dunes Solar Energy Project', available at http://www.basinandrangewatch.org/CrescentDune.html (accessed on 10 July 2017). The Crescent Dunes plant, which does not have dry cooling consumes applied for a permit to dispose of waste water that speaks for itself:

> August 2015—Tonopah NV—The Crescent Dunes Solar Energy Project is applying for a five year permit to discharge up to 0.5 million gallons per day of industrial process wastewater, including cooling tower and boiler blowdown water, reverse osmosis reject water, demineralized water, oil/water separator water, and floor drain water, to three double-lined evaporation ponds. The permit would also authorize the discharge of up to 0.2 million gallons per day of demineralized water for heliostat washing, and 0.2 million gallons per day of untreated water for dust control.

18   Susan Kraemer, 'One Weird Trick Prevents Bird Deaths at Solar Towers', *CleanTechnica* (16 April 2015), available at http://cleantechnica.com/2015/04/16/one-weird-trick-prevents-bird-deaths-solar-towers/ (accessed on 10 July 2017).

19   According to the American Bird Conservancy, crashing into reflective glass windows that they mistake for the sky kills close to a billion birds a year in the United States, see http://abcbirds.org/program/glass-collisions/ (accessed on 10 July 2017).

20   Basin and Range Watch, 'Crescent Dunes Solar Energy Project'.

21   See https://www.wired.com/2016/05/huge-solar-plant-caught-fire-thats-least-problems/ (accessed on 10 July 2017).

22   See http://world-nuclear.org/information-library/current-and-future-generation/electricity-and-energy-storage.aspx (accessed on 9 February 2017).

23   See https://electrek.co/2016/11/01/breakdown-raw-materials-tesla-batteries-possiblebottleneck/ (accessed on 15 September 2017).

24   See http://www.infomine.com/investmandent/metal-prices/coal/ (accessed on 10 July 2017). In January 2017, standard American coal costs $40–44 a tonne while oil costs $53 a barrel or close to $400 a tonne. A tonne of oil has 1.42 times as much energy as a tonne of standard coal.

25   David Levitan, 'Republicans Offer to Tax Carbon Emissions', *Scientific American* (8 February 2017), available at https://www.scientificamerican.com/article/republicans-offer-to-tax-carbon-emissions/ (accessed on 10 July 2017).

26   See http://www.titantracker.com/v_portal/apartados/apartado.asp?te=686 (accessed on 10 June 2015) and http://www.slideshare.net/titantracker/titan-tracker-csp-heliostat-design (accessed on 23 January 2017).

27   Ibid.

28   Forty four per cent efficiency against the conventional best of 38 per cent. See http://www.brighthubengineering.com/power-plants/32896-how-are-supercritical-boilers-different-from-subcritical-boilers/ (accessed on 10 July 2017).

29   See note 26.

30   Ibid. See graph of pointing accuracy in paper.

31 Torresol has published the precise specifications of the Gemasolar
    plant. The operational assumptions on which this estimate is based
    are that heliostats cost $120/m²; a permanent labour force of 90,
    against Gemasolar's 45; an average salary of $6,000 per year (about
    a fifth of the European average), but an interest rate of 7 per cent
    instead of 3 to 5 per cent in Europe and the United States and a
    plant life of 30, not 40 years. It also assumes an annual operation and
    maintenance cost of 2 per cent of the capital cost and adds 0.67 cent
    of capital cost to the annual operation costs to cover land lease and
    incidental costs not included above.

32 Heinz-Muller Steinhagen, 'Concentrating Solar Thermal Power'.

33 This tallies fairly closely with the estimates of the Sandia National
    Laboratories, adjusted to 2014 prices. There is now so much data
    available on CSP that it is possible to make fairly precise estimates of
    its cost of production under a range of assumptions about the future
    cost of heliostats. This was done by the Sandia National Laboratories
    at 2006 prices and again by Titan Trackers using their patented track-
    ing system in 2010. Based on these, it is assumed here that the helio-
    stats have a life of 30 years (amortisation rate of 3.33 per cent) that
    average annual expenditure on operation and maintenance amounts
    is 2 per cent of the capital costs and an interest rate of 7 per cent
    instead of 3 to 5 per cent in Europe and the United States. It also adds
    0.67 cent of capital cost to the annual operation costs to cover land
    lease and incidental costs not included above.

34 See https://www.theguardian.com/environment/2008/jul/23/solar-
    power.windpower (accessed on 10 July 2017).

35 See https://www.theguardian.com/environment/2016/feb/04/
    morocco-to-switch-on-first-phase-of-worlds-largest-solar-plant
    (accessed on 10 July 2017).

36 Peter Engelke, Russell Sticklor, 'Water Wars: The Next Great Driver
    of Global Conflict?' *The National Interest* (15 September 2015),
    available at http://nationalinterest.org/feature/water-wars-the-next-
    great-driver-global-conflict-13842 (accessed on 23 January 2017).

37 Dun & Bradstreet, 'India's Energy Sector: Executive Overview', avail-
    able at http://www.dnb.co.in/IndiasEnergySector/Power_Gen.asp
    (accessed on 10 July 2017).

[38]   Utpal Bhaskar, '16-year Delays, 2,532% Cost Overruns Shock House Panel', *Livemint* (12 April 2007).

[39]   See http://www.dnb.co.in/IndiasEnergySector/images/Table3.8.gif (accessed on 10 July 2017).

[40]   Data for 'overnight' capital costs obtained from US Energy Information Administration: Updated Capital Cost Estimates for Utility Scale Electricity Generating Plants. These have been adjusted to obtain estimates for India (and China) in the ways described in the table's footnotes.

Chapter Five

The Strange Neglect of Biomass

Coal- and gas-based power generation are not the only sources of greenhouse gases. Finding alternatives to petroleum-based transport fuels is almost equally important, because it is where the world will start running out of raw materials first. This makes the neglect of biomass, which is the only alternative feedstock for transport fuels, hard to understand. Technologies for extracting fuel gas from biomass have been known for more than a century, but these have been systematically shunted aside in favour of so-called 'advanced' biofuels whose supply cannot possibly be increased sufficiently in the time we have left.

The Global Apollo Programme's proposal to concentrate future research upon the development of 'economically viable' technologies that can replace fossil fuels with renewables by 2025 is the first unequivocal admission by an eminent body of policy makers in the West that the battle against climate change cannot be left to market forces alone.

This change of approach brings economics and technology back into the search for solutions, but it has so far done so only lopsidedly, for it has been preoccupied almost exclusively with the generation of electricity and process heat, to the near-complete exclusion of transport fuels. But oil and gas are where the world will first run out of non-renewable resources, so it is where the collapse of the global economic system predicted by the Club of Rome will begin. It is, therefore, even more, important to find and deploy technologies that can avert the struggle over depleting reserves of oil and gas.

The only primary source of energy that can replace oil and gas is biomass, which is the oldest fuel on earth. Even today, it meets a tenth of human energy needs, as the cooking and heating fuel of the poor in traditional societies. But, inexplicably, as the search for alternatives to fossil fuels has grown more urgent, biomass has slipped further and further into obscurity. Today, the most ardent advocates of renewable energy seldom mention biomass.

The false promise of ethanol

Every rule, however, has an exception, and the exception in this case is ethanol. Ever since Brazil pioneered its use in 1978 as an octane-boosting agent for blending with gasoline the only biofuel that policy makers in most countries have actively promoted is ethanol. Between 2000 and 2011, the world's production of ethanol fuel went up by five times from 4.4 billion to more than 21 billion gallons. The United States accounted for 62 per cent of this output (13.2 billion gallons). Brazil came second with 5.25 billion gallons. Just these two countries, therefore, produced 87 per cent of all the ethanol used as a transport fuel in the world in that year. Today, ethanol meets 18 per cent of Brazil's transport fuel needs.

Brazil's leaders have been lauded for their far-sightedness in developing ethanol as a transport fuel. The country is considered to have the world's first sustainable biofuels economy, because in 1993 it mandated (but has not quite achieved) a 22 per cent admixture of ethanol with gasoline and because the ethanol component produces barely 10 to 12 per cent as much CO_2 as the gasoline it replaces.

Not surprisingly, when President George W. Bush launched a new energy policy in his Energy Security and Independence Act,

2007, he did it with a 51 cents a gallon subsidy on the ethanol that was blended with gasoline. He underlined the Brazil connection by announcing his programme just one day after President Lula had announced a $6 billion programme to increase the production of sugar cane for ethanol in Brazil. In conscious emulation, Bush set a target of 20 per cent admixture of ethanol with gasoline by 2017.[1]

The Bush–Lula initiative was hailed by environmentalists as a decisive first step towards saving the planet. But neither President had that in mind when they took it. Brazil pioneered the use of ethanol as a transport fuel in 1976 because it saw, in the production of ethanol as a transport fuel, an opportunity to kill two birds with one stone. Brazil was then, and remains today, by far the world's largest exporter of sugar. But the price sugar fetches in the world market fluctuated hugely from year to year because sugar cane takes 12 to 14 months to grow. As a result, when farmers plant their cane they have to guess what prices will be a year later. If, anticipating high prices, they increase their acreage it results in a glut and a sharp fall in world prices, to which the only possible response is to reduce the acreage under cane the next year. This begins the next phase of the cycle. When oil prices shot up, the Brazilian government saw an opportunity to kill two birds with one stone: By using a part of the sugar output to produce ethanol and changing the amount in response to sugar price fluctuations, it could stabilise the export price of its sugar and save valuable foreign exchange on the import of oil, at the same time.

The allocation of sugar cane to ethanol and sugar production is therefore decided each year by cane millers based on expected sugar and ethanol prices and market demand. In 2010, 55 per cent of the crop went to ethanol. But in the very next year, when a drought reduced the output of sugar and pushed its price sky-high, the millers raised the allocation of cane to the production of sugar by so much that Brazil found itself importing ethanol from the United States to make good its deficit.[2]

Averting climate change was also not the first thing on President Bush's mind when he announced his ethanol programme.

His purpose, as summed up by David Jackson of *USA Today* was as follows:

> First, expanding world ethanol markets would help the USA and other nations reduce their dependence on foreign oil. Also, reduced demand for oil might reduce the clout of Venezuelan President Hugo Chávez, who has tried to use his nation's oil reserves to undercut U.S. policies in the region.

Knowing that the United States could not by itself produce all the ethanol it needed, Bush decided to forge a partnership with Brazil. The high point of his visit to Brazil in March 2007, a month after he and President Lula had announced their respective plans for increasing the production of ethanol, was therefore the signature of an 'Ethanol Partnership' under which Brazil would increase its export of ethanol from 3 billion litres in 2006 to 200 billion litres in 2025.[3]

The US–Brazil ethanol partnership is now very largely a history. Eight years after it was signed, the share of ethanol blended with gasoline in the United States is stuck at 10 per cent.[4] As is described in a later chapter, the reason is the failure to produce cellulosic ethanol (ethanol from fibrous biomass as distinct from sugar and starch) on any significant scale. This has forced it to rely on corn to produce it. But ethanol produced from food crops has one overwhelming drawback. Even when produced in relatively small quantities it has the capacity to disrupt not only the ecosystem but spread political turmoil across the world. In the first year after President George W. Bush launched his ethanol programme, 20 per cent of the American corn crop was diverted from feeding human beings and cattle to producing ethanol. This triggered a chain reaction around the globe that had still not ended in 2011 and may partly have been responsible for the turmoil of the Arab Spring. In September 2007, barely six months after the start of the ethanol programme, the *New York Times* reported:

> The distortions in agricultural production are startling. Corn prices are up about 50 percent from last year, while soybean

prices are projected to rise up to 30 percent in the coming year, as farmers have replaced soy with corn in their fields. The increasing cost of animal feed is raising the prices of dairy and poultry products.

The news from the rest of the world is little better. Ethanol production in the United States and other countries, combined with bad weather and rising demand for animal feed in China, has helped push global grain prices to their highest levels in at least a decade. Earlier this year, rising prices of corn imports from the United States triggered mass protests in Mexico. The chief of the United Nations Food and Agriculture Organization (FAO) has warned that rising food prices around the world have threatened social unrest in developing countries.[5]

Six months later, *Time* magazine reported how this had destabilised political systems around the world by pushing up food prices. 'By diverting grain and oil seed crops from dinner plates to fuel tanks, biofuels are jacking up world food prices and endangering the hungry.' The grain it takes to fill an SUV tank with ethanol could feed a person for a year. Harvests are being plucked to fuel our cars instead of ourselves. The UN's World Food Programme says it needs $500 million in additional funding and supplies, calling the rising costs for food nothing less than a global emergency. Soaring corn prices have sparked tortilla riots in Mexico City, and skyrocketing flour prices have destabilised Pakistan, which was not exactly tranquil when flour was affordable. In Brazil, for instance, only a tiny portion of the Amazon is being torn down to grow the sugar cane that fuels most Brazilian cars. More deforestation results from a chain reaction so vast it is subtle: US farmers are selling one-fifth of their corn to ethanol production, so US soya bean farmers are switching to corn, so Brazilian soya bean farmers are expanding into cattle pastures, so Brazilian cattlemen are displaced to the Amazon. Such is the remorseless economics of commodity markets. 'The price of soybeans goes up', laments Sandro Menezes, a biologist with Conservation International in Brazil, 'and the forest comes down'.[6]

A report prepared by the World Bank in April 2008 concluded that 75 per cent of the increase in food prices since 2006 had been caused by the diversion of corn and soya beans into the production of ethanol and biodiesel, and that only 3 per cent could be attributed to higher consumption in China and India. This was suppressed by the bank on the grounds that it overemphasised the role of diversion to the fuel sector and did not sufficiently take into account a number of other causes including inclement weather, but this only weakens its overall conclusion without invalidating it.

To cap it all, the production of ethanol from corn is actually increasing the emission of GHGs. A study by the US Environmental Protection Agency of the life cycle impact of the switch to ethanol showed that over a 30-year period, the technologies currently in use will increase the emission of GHGs. It is only over a hundred years that the balance becomes favourable. Unfortunately the world no longer has a hundred years in which GHG emissions can be brought down.[7]

The strange neglect of methanol

There is, however, another transport fuel that can be produced almost as easily from biomass as ethanol is produced from starch, but does not take food from people's mouths. This is methanol. It is one of the most widely used and versatile chemicals on the planet. It is the basic ingredient of paints, solvents, plastics and synthetic fibres. It is a crucial feedstock in the production of a wide variety of chemicals such as formaldehyde, and petrochemicals, and it is an outstanding transport fuel, with all of the flame properties as ethanol. Today, there are more than 90 plants all over the globe, which produce 75 million metric tonnes of methanol. More than 100,000 tonnes a day, almost half of the total output, is being used as a petrochemical intermediate or as a transport fuel. Methanol has also been identified as the ideal hydrogen generator for fuel cells.

Methanol is currently being produced from natural gas and so does not qualify as a biofuel, but it can be produced from biomass almost as easily as ethanol. In its pure form it has an octane rating of 122. Since the methanol molecule contains its own oxygen, it has a much higher flame speed than gasoline. As a result, automotive engines running on pure or 85 per cent methanol (an extensively tested blend) can be designed to have compression ratios as high as 18 to 1, against 8 to 1 for unleaded gasoline, and engine speeds of 12,000–14,000 revolutions per minute, against a safe maximum of 6,000–7,000 rpm for gasoline engines. As a result, with less than half the energy content of gasoline, pure methanol can deliver up to three quarters of its propulsive power.

These are qualities it shares with ethanol, but it has one inestimable advantage over ethanol: it can be produced from literally any kind of biomass waste, be it sewage sludge, municipal solid waste (including plastics), or any variety of agricultural residue. Therefore, not only does it cause no disruption of the food chain, and therefore of food prices, but its symbiotic relationship with food production ensures that the more the world demands transport fuels, the more food will there be for all to eat.

Methanol has been used as a transport fuel for much longer than ethanol. While Brazil pioneered the latter in 1978, the German government pioneered the large-scale use of methanol in 1943, during the Second World War. Methanol was the mandated fuel for most motor car races from 1965 till 2008 (when it was replaced in some major races by ethanol). The automobile race associations took this step because unlike gasoline, methanol does not explode when provided with a spark in the open air, but only burns. This quality was noticed at the Indianapolis 500 motor car race in 1963 when a two car crash and explosion killed two of the world's most famous drivers and more than 130 bystanders. A third car also piled into the first two but did not explode. As a result the driver was able to walk away. The third car had been running on methanol.

Outside the race track, passenger cars have been running on blends of methanol with gasoline in South Africa since the 1950s. They were running on up to 85 per cent methanol in California from the early 1990s till as recently as 2006. Why then is the world ignoring this extraordinary fuel? Whenever one raises this question at conferences or discussions, the response is usually 'isn't methanol toxic to humans?' The answer is that, if we ingest more than 0.2 litres—about two wine glasses—it will first blind and then kill us. But since no one is proposing methanol as a beverage, this objection makes very little sense.

Side by side with the unthinking acceptance of this objection is a profound ignorance of its many virtues: that the exhaust from burning methanol even in low compression, dual fuel vehicles, contains no nitrous and sulphurous oxide gases, no methane, virtually no carbon monoxide (CO) and only traces of formaldehyde and carbonyls that can easily be filtered out, leaving only carbon dioxide and steam;[8] and that methanol is also not a pollutant. An oil spill of the Exxon Valdez variety would pose no environmental threat because methanol dissolves within minutes and breaks down within days. Lastly, using bio-methanol as a transport fuel could actually extract some CO_2 from the air because while its production from biomass is carbon-neutral, its much greater efficiency in combustion reduces the amount of carbon needed for the same travel distance.

Methanol has been considered as an alternative, or additive, to gasoline in passenger cars since the very first oil shock of the early 1970s. In 1973, shortly after the first oil price shock, two MIT scientists Reed and Lerner started a project to test the characteristics of methanol as a transport fuel. Blending different amounts with gasoline, they had found in initial tests that adding up to 20 per cent of methanol improved the fuel efficiency of an automobile, before MIT abruptly withdrew the funding from their project.[9]

However, the idea did not die. In 1978, the Government of New Zealand took a look at its abundant natural forests of *pinus radiata* and drew up a plan to replace gasoline and diesel with methanol and a close substitute, dimethyl ether (DME), obtained by gasifying the wood residues from its sylviculture programme and converting the resulting synthesis gas into these transport fuels. The plan was ready to implement in 1981 but was put it into cold storage when oil prices crashed in that year.

Four years later, in 1985, two scientists Charles Gray and Jeffrey Allison published a book *Moving America to Methanol*, in which they pointed out that the United States (and all rich northern countries) could solve several problems at once by moving their transport fuel base from fossil fuels to methanol. Their primary concern then was with acid rain and related air pollution problems. These would disappear because the exhaust from its combustion yielded virtually nothing but carbon dioxide and water vapour. They also pointed out that this would free the United States to a considerable extent from its growing need to import oil.[10]

In the United States, there was a sharp spike of interest in methanol after the federal government banned the addition of lead to gasoline in 1976. Two scholars from the Sloan School of Management at MIT, L. Bromberg and W.K. Cheng, have given a detailed description of the experiments that were conducted between 1980 and 2005: The initial interest in methanol was not as an alternative to gasoline but as an additive that would raise its octane rating, after the addition of lead was banned in 1976. This early interest in methanol spawned several programmes, mainly in California. An experimental programme ran from 1980 to 1990 for the conversion of gasoline vehicles to an 85 per cent methanol and 15 per cent gasoline blends. Gasoline-driven vehicles were modified to get the best results from this blend. The tests showed that these dedicated methanol (M-85) vehicles performed as well or better than their gasoline counterparts. The vehicles used in the initial programme were provided by US automakers, who were subsidised to produce a fleet of vehicles for use mainly in California. Ten automakers participated, producing 16 different models, from light-duty vehicles to vans and even heavy-duty vehicles (buses) with over 900 vehicles. The automakers provided spark-ignited engines and vehicles that were well engineered, which addressed issues with methanol compatibility.

> One of the fleets, with about 40 gasoline based and methanol-based vehicles (for direct comparison), was operated by the US Department of the Environment's laboratories from 1986–1991. Both the baseline gasoline and retrofitted M85 (85 percent methanol, 15 percent gasoline) vehicles were rigorously maintained, with records to determine their performance.... These (M-85) vehicles performed as well or better than their gasoline counterparts with comparable emissions, which was a bonus, since methanol emissions were less prone to ozone formation. The test drivers also noted that the acceleration of the M-85 vehicles was better: the same vehicles took one second less to accelerate from 0 to 100 kmh.

> But since there were very few gas stations that provided this blend, and the engines would not run on pure gasoline, they did not prove popular with the users. An evaluation report for the Program, therefore, concluded that 'the result [was] a technically sound system that ... frustrated drivers trying to get fuel, generating an understandably negative response to the operator'.[11]

Upon the completion of the dedicated vehicle programme in 1990, fleets with flexi-fuel engines were tested, mostly in California. Ford build 705 cars with flexi-fuel engines, ranging from the 1.6 litre Escort, to the 3.0 litre Taurus and the 5.0 litre Crown Victoria. There were also a few 5 litre Econoline vans. The trials were so successful that automakers began offering flexible-fuel vehicles (FFVs) to customers from 1992. The number of M-85 FFVs in the United States peaked in 1997 at just over 21,000. Of these, 15,000 were in California, which had more than a hundred public and private refuelling stations. At the same time, there were hundreds of methanol-fuelled transit and school buses.

But the farm lobby was not to be denied. Ethanol eventually displaced methanol in the United States. In 2005, California stopped the use of methanol after 25 years and 200 million miles of operation. In 1993, at the peak of the programme, over 12 million gallons of methanol were used as a transportation fuel. In addition to California, New York state also road-tested a fleet of methanol-fuelled vehicles, with refuelling stations located along the New York Thruway. A US Department of Energy report concluded that methanol had lost out to ethanol because it did not have any organised lobby behind it. Bromberg and Chen concluded nonetheless that: 'While methanol has not become a substantial transportation fuel in US ... the methanol FFVs have demonstrated that it is a viable fuel and (that) the technology exists for both vehicle application and fuel distribution'. It ascribed the neglect of methanol to the fact that, unlike ethanol, which has been promoted by and the farm lobby in the United States and the Brazilian government, there is no lobby-pushing methanol. But there is evidence that this could have been because from the early 1970s, the oil majors sensed a threat in methanol that they did not in ethanol, because of the virtually inexhaustible supply of biomass residues from which it could be made. Ethanol, therefore, failed the sixth test of acceptability outlined above, of not posing a threat to those who are already heavily invested in existing energy technologies.[12]

Methanol also failed to become a substantial component of transport fuel in the global economy because of sheer bad timing. First, the trial programme was begun in the 1980s at the beginning of a prolonged period of falling real prices of oil and oil products and came to an end in 2005 just when the fall in oil price reversed itself abruptly. In 2006, therefore, there was no lobby in the United States, or anywhere else in the world, advocating blending methanol with gasoline. Then a number of commercial plants designed to produce methanol from black liquor, a toxic effluent of the paper industry, were shelved because of the onset of global recession. This is described in greater detail later in the book.

George W. Bush's costly mistake

In 2006, when President Bush was putting the final touches to his Energy Security Bill, he had a clear choice between ethanol and methanol. On one side was the powerful farm lobby. On the other was a group of scientists and officials in the Department of Energy who were quietly but vigorously pushing for methanol. All the scientific arguments were in favour of methanol. But the political ones favoured ethanol and ethanol won. Hindsight will one day show that this was a costly mistake. For if the world has wasted 23 years since the Rio Earth Summit doing next to nothing to avert the growing threat from global warming, Bush is responsible for 9 of those years. For not only has the technology for producing cellulosic ethanol not been proven, but it is never going to be an efficient source of biofuels to start with.

Commending an article 'Beyond Corn: Ethanol's Next Generation' from the *Chicago Tribune*,[13] Robert Rapier, a chemical engineer and noted writer on alternate energy technologies, presented the case for methanol succinctly on the energy website 'The Oil Drum—Discussions About Energy and the Future' in 2006:

> More money than ever before is being poured into cellulosic ethanol, but there are multiple hurdles that have proven difficult to overcome…. The first is that plants have evolved defence mechanisms to prevent the cellulose from being easily broken down. Cellulose is actually a polymer—a long chain of connected sugars, and it is intermingled with hemicellulose and lignin. Cellulose provides structural strength to the plant walls. If it was easily broken down, microorganisms could attack the plants and limit their structural stability. What this means is that the cellulose must first be broken down with steam or a strong acid into component sugars that can be fermented, and this adds to the production costs. It is primarily this step that differentiates cellulosic ethanol from grain or sugarcane ethanol.
>
> The second hurdle is common to all ethanol fermentation processes, but not to gasification processes. The ethanol that

is produced in a fermentation process is highly diluted with water. In fact, the ethanol produced from fermenting grain typically makes up only 15–20% of the solution, with the remainder being mostly water. For cellulosic ethanol, the picture is much worse. The crude ethanol in this case is typically less than 5%, with the remainder being water. Separating water and ethanol (therefore becomes) a very energy-intensive process....

'Biomass gasification', Rapier went on,

is different from cellulosic ethanol in at least two major respects. First of all, it is a combustion, not a fermentation process. As a combustion process, it can be self-sustaining once the combustion is initiated. It does not require continual inputs of energy as is the case with a fermentation process. The products of biomass gasification are syngas and heat, if the reaction is operated in an oxygen-deficient mode or CO_2 and steam (and much more heat) in the case where sufficient oxygen is supplied. In the case of the former, the syngas can be further reacted to make a wide variety of compounds, including methanol, ethanol or diesel (via the Fischer–Tropsch reaction). A biomass gasification process followed by conversion to a liquid fuel is commonly referred to as a biomass-to-liquids (BTL) process.

However, there is one other major factor that differentiates biomass gasification from cellulosic ethanol.... In the (manufacture of) of cellulosic ethanol, only the cellulose and hemi-cellulose are partially converted after being broken down to sugars. The lignin and other unconverted carbon compounds end up as (wet) waste, suitable for burning as process fuel only if thoroughly dried. Conversion is limited to those components which can be broken down into the right kind of sugars and fermented.

Gasification, on the other hand, converts all of the carbon compounds. Lignin, a serious impediment and waste product in the case of cellulosic ethanol, is easily converted to syngas in a gasifier. The conversion of carbon compounds in a gasification process can be driven essentially to completion if desired, and the resulting inorganic mineral wastes can be returned to the soil.[14]

Lignin typically makes up about 25 per cent of the dry matter in any type of biomass. Rapier was, therefore, pointing out that even if the hemi-cellulose is broken down completely, cellulosic ethanol will transform only three quarters of the biomass into ethanol.

But this debate ended when it became apparent that the prime mover behind the promotion of ethanol was America's powerful farm lobby. As a result in March 2007, the US government awarded $385 million in grants to six projects for the production of cellulosic ethanol from wood chips, switch grass and lemon peels using enzymes to turn the cellulose and hemi-cellulose into ethanol. Five years later, as of April 2012, not a single plant had commenced production. One of the companies Range Fuels actually constructed a plant to produce 100 million gallons of ethanol a year in 2007–2008, but shut it down in January 2011 without producing a single gallon. By then the company had received $162 million in grants and guaranteed loans. A second company BlueFire Renewables (now Bluefire Renewables Inc.) had still to get off the ground on 25 April 2012 and was describing itself as a 'development stage company'.[15] None of the other recipients had established a functioning cellulosic ethanol plant.[16]

Had the $385 billion been given to companies for the production of methanol or other transport fuels that can be produced by gasifying biomass instead of trying to ferment it, the United States would have been well on the way to substituting biofuels for fossil fuels in the transport sector, and the rest of the world would have been following suit.

Why ethanol continues to beckon

Why, despite this failure, does cellulosic ethanol still beckon investors? The reason is that it meets five of the six criteria for acceptability that were outlined in Chapter 3. First, the technology for its manufacture from food crops, such as sugar, corn and molasses, is fully proven. Second, its manufacture is phenomenally profitable.

Third, replacing gasoline with ethanol does not disrupt the existing fuel supply infrastructure.

It blends easily with gasoline up to 20 per cent, and blends of up to 85 per cent ethanol can be transported and stored safely with only minor changes in the supply infrastructure. At the users' end, flexi-fuel engines have already been designed that can run on every blend from all gasoline to all ethanol. So shifting the transport fuel base of society to it will impose next to no external costs upon society.

Finally, the oil industry in particular knows that a technological breakthrough that allows cellulosic ethanol to be manufactured from virtually any type of biomass is still far away. Even if a 20 per cent ethanol blend becomes possible, it will not dent its bottom line. So none of its investments will go to waste. It has also done its sums and knows that much higher blends are simply not feasible, for want of ethanol. Brazil's ambitious ethanol plan only helps to underscore this point. To increase its ethanol exports from 750 million US gallons in 2006 to 50 billion gallons in 2025, Brazil would have to bring another 24 million hectares of land under sugar cane. But even that would replace barely a quarter of the gasoline that the United States consumes today. The question of replacing not just gasoline, but high speed diesel does not therefore arise. This accounts for the absence of opposition from the oil and gas lobby.[17]

Notes and references

1   Isabella Kenfield, 'Brazil's Ethanol Program Breeds Rural Poverty, Environmental Degradation', Global Research (8 March 2007), available at http://www.globalresearch.ca/index.php?context=va& aid=5012 (accessed on 10 July 2017). President Lula announced that he would increase federal funding for Brazil's from 22 January 2007. President Bush announced his ethanol plan in his State of the Union message on 23 January 2007.

2   Farmgateblog.com, 'Will Brazilian Ethanol Really Compete with US Ethanol', available at http://www.farmgateblog.com/article/

will-brazilian-ethanol-really-compete-with-us-ethanol (accessed on 10 July 2017).

[3] David Jackson, 'US, Brazil Plan Ethanol Partnership', *USA Today* (2 March 2007), available at http://usatoday30.usatoday.com/news/world/2007-03-01-us-brazil-ethanol_x.htm (accessed on 10 July 2017).

[4] Matthew L. Wald, 'For First Time, E.P.A. Proposes Reducing Ethanol Requirement for Gas Mix', *The New York Times* (15 November 2013), available at http://www.nytimes.com/2013/11/16/us/for-first-time-epa-proposes-reducing-ethanol-requirement-for-gas-mix.html?_r=0 (accessed on 21 February 2014).

[5] New York Times Editorial, 'The High Costs of Ethanol', *The New York Times* (19 September 2007), available at http://www.nytimes.com/2007/09/19/opinion/19wed1.html (accessed on 10 July 2017).

[6] Michael Grunwald, 'The Clean Energy Scam', *Time* (27 March 2008).

[7] EPA, 'Lifecycle Analysis of Greenhouse Gas Emissions from Renewable Fuels' (May 2009), available at http://www.epa.gov/otaq/renewablefuels/420f09024.pdf (accessed on 10 July 2017).

[8] A 2015 study of emissions from using pure methanol in dual fuel (and therefore low compression) automobile engines, by the Beijing Institute of Technology showed that compared to gasoline, methanol emitted 65 per cent less particulates, 11 per cent less CO, 8 per cent less CO_2, but 39 and 19 per cent more formaldehyde and carbonyls, respectively. X Wang, Y Ge, L Liu and H Gong, 'Regulated, Carbonyl Emissions and Particulate Matter from a Dual-fuel Passenger Car Burning Neat Methanol and Gasoline', *SAE Technical Paper* 2015-01-1082 (2015), doi:10.4271/2015-01-1082, available at http://papers.sae.org/2015-01-1082/ (accessed on 10 July 2017).
Emissions from methanol would, however, fall sharply if the compression ratio were raised to the optimum levels for methanol, which could be as high as 18 to 1.

[9] A fuller description of the project and how it got scuttled is given in Chapter 9.

[10] Gray and Allison were not, then, thinking of global warming. They felt no qualms about, and indeed recommended, that the methanol should be obtained by gasifying coal, which America had in abundance. Today coal gasification to produce methanol would only

compound the problem of global warming because it would almost double the total emission of CO_2 gasoline-equivalent gallon of gas.

[11] Bromberg, L. and Cheng, W.K., 'Methanol as an alternative transportation fuel in the US: Options for sustainable and/or energy-secure transportation', Sloan Automotive Laboratory, Massachusetts Institute of Technology, (2010). National academy of sciences: https://trid.trb.org/view.aspx?id=1098776

[12] As is described in Chapter 6, this is not just the oil industry.

[13] Michael O'neal, 'Beyond Corn: Ethanol's Next Generation'; 'Scientists Seek Cheap, Plentiful, Energy Alternatives', *Chicago Tribune* (13 October 2006), available at http://www.chicagotribune.com/news/nationworld/chi-0610130128oct13,0,2156857.story (accessed on 10 July 2017).

[14] Robert Rapier, 'Cellulosic Ethanol vs. Biomass Gasification', *Energy Trends Insider* (22 October 2006), available at http://www.energy-trendsinsider.com/2006/10/22/cellulosic-ethanol-vs-biomass-gasification/ (accessed on 10 July 2017).

[15] *The New York Times*, Business Day section, Company Information press release (25 April 2012).

[16] Dirk Lammers, 'Gasification May Be the Key to US Ethanol', *CBS News* (4 March 2007), available at http://web.archive.org/web/20071122145559/ http://www.cbsnews.com/stories/2007/03/04/ap/tech/mainD8NLISJ80.shtml (accessed on 10 July 2017).

[17] This is almost certainly why James Hansen concedes, although with regret, that there is no immediate prospect of replacing more than a fraction of fossil-based transport fuels with biofuels and urges governments to concentrate on phasing out all coal-based power plants that do not sequester the CO_2 they produce by 2030 at the latest.

Chapter Six

Transport Fuels Are the Place to Start

The technology that can unleash the power of biomass is not fermentation but gasification. This chapter describes how hurdles to harnessing it to convert almost any type of biomass into transport fuels have been crossed.

Ethanol is not the answer to mankind's prayers, but its ready accept-ability has shown that the battle against climate change could have been, and indeed should have, started not in the power but in the transport sector. This is in direct opposition to the urging of James Hansen and other noted climatologists, of advocacy groups such as Greenpeace and the Sierra Club, and most recently of the Global Apollo Programme that the starting point must be the closure of coal-based power plants.[1] The reason why Hansen and others are wrong is not that transport generates more GHGs than power gen-eration but that transport relies on oil and natural gas, the fossil fuels whose supply is most limited, and whose prices (per unit of energy) are not only the highest per unit of energy contained but also the most prone to speculative surges in the world today. Investors in new energy technologies will need to make a profit, and the profit from replacing oil and gas with biofuels will be sev-eral times greater than by replacing coal in the power sector.[2]

Gasification

Is there a technology that can replace oil and natural gas with bio-fuels in the next 30 or 40 years? The answer, as chemistry Nobel Laureate George Olah explained at length in 2007, is gasification, also known, more generically, as destructive distillation.[3] Gasification is the combustion of any organic matter in a limited supply of air, and therefore of oxygen. Human beings first used a variant of this technology 30,000 years ago to make charcoal. Today, refined a thousandfold, it is one of the important keys to averting catastrophic climate change.

While the complete combustion of biomass yields only large quantities of carbon dioxide, incomplete combustion yields hydrogen, CO and a far smaller amount of CO_2. At low temperatures it also yields heavier gases such as methane, butane and pentane, tarry oils and tar, but these also break up into CO and hydrogen as the gasification temperature is raised.

A mixture of CO and hydrogen is called 'synthesis gas' or 'syngas'. This is the philosopher's stone of the modern chemicals industry, for it is the basic building block of the production of virtually the entire range of petrochemicals from fertilisers, to plastics and synthetic fibres. Synthesis gas is the essential prerequisite for producing ammonia which, when combined with nitrogen, yields urea, the most widely used chemical fertiliser in the world. It is also the starting point of all forms of petrochemicals. Another widely used product of gasification is methanol. But methanol is only one of the transport fuels that can be produced from syngas. The chemical reaction that converts the syngas into transport fuels and other liquid petrochemicals is the *Fischer–Tropsch synthesis.* Using different temperatures, pressures and catalysts, it yields all of the above.

Several plants have produced, and many continue to produce, transport fuels and synthetics from coal in Canada, the United States, Malaysia, Qatar, South Africa, Finland, China and India. Obtaining syngas from biomass has several advantages: it contains almost no sulphur—a pollutant that destroys catalysts; natural gas does not have to be cracked—a fairly expensive process. Best of all, producing transport fuels from biomass is carbon-neutral, for the CO_2 released from the combustion of biofuels has been sucked in by the biomass from the air.

The immense potential of biomass gasification was succinctly described in a report presented to the Executive Committee of the International Energy Association at a meeting in Golden, Colorado, in 2007. The research team concluded that 'almost all types of biomass with less than 50 per cent moisture can be gasified with air, enriched air, or oxygen to produce a fairly uniform-in-quality fuel that could readily substitute for fossil fuels and fossil fuel derived products'. It did, however, caution the committee members that what this could only be done in large modern gasifiers and not in the small, air-driven local and village biomass gasifiers that were being experimented with in India, China and elsewhere.[4]

Gasification is often confused with an older but similar process called pyrolysis. But the two are very different chemical reactions. In pyrolysis, heat has to be provided from outside to raise the temperature of the organic matter to the level at which it starts releasing these gases. It is therefore what chemists call an 'endothermic' reaction, one that requires energy from the outside. Because of the presence of a limited amount of oxygen, gasification generates its own heat and is therefore an 'exothermic' reaction. This is especially true when pure oxygen is used in place of air. Temperatures then reach well over 1,000°C.

The basic procedure is to stack the biomass in a column, set fire to it at the bottom, keep the fire burning there by pumping in required quantities of air or oxygen and allow the heat from the combustion at the bottom to filter up and leach out the carbon-rich gases from the upper layers. As each layer finishes burning, the ash passes out and the entire column moves down, whereupon the next layer also starts burning. By then all the hydrogen, and a large part of the carbon in the biomass, has been released as CO.

While pyrolysis usually takes place at temperatures below 500°C, gasification reactions are carried out at temperatures from 800° to 3,000°C. Most current gasifiers operate in the neighbourhood of 1,000°C. The higher the temperature, the lower is the yield of the heavier gases that have to be broken down further, till at temperatures above 1,000°C, one is left with only hydrogen, CO and CO_2.

A brief history

The first organic matter to be gasified on a commercial scale was coal. This was done by the German government during the Second World War to produce transport fuels for their war machine. After the war, the South African conglomerate South Africa Synthetic Oil Liquid (SASOL) developed the technology further to produce a wide variety of transport fuels and petrochemicals. Today, SASOL meets close to half of South Africa's transport fuel needs from coal gasification. This technology is now being adopted on a large scale in China and is being explored seriously by the Indian government.

Scientists became interested in gasifying other forms of biomass to produce transport fuels only after the first oil price shock in the 1970s. The first experiments replaced coal with wood, but the technology has now developed to the point where any form of organic matter, ranging from sewage sludge and MSW to crop residues, paper, rubber, plastics and wood wastes, can be successfully converted into 'synthesis gas' and thence to transport fuels. To do this, the producer gas is cooled, cleaned, purged of a part of the CO_2 and residual methane if there is any and put through what is known as the water gas shift reaction to adjust the proportion of hydrogen and CO to the desired ratio and fed into the synthesiser for the Fischer–Tropsch synthesis to work its magic. This entire downstream process is well known and has been in widespread use for more than a hundred years.

Starting with biomass instead of oil or gas, however, changes the proportions of the four gases, namely CO_2, CO, hydrogen (H_2)

and methane (CH_4) in the producer gas. What is more, the proportions vary from one type of biomass to another and from one season of the year to the next. It is not surprising, therefore, that the single most important goal of all research into biomass gasification has been to find cheaper, more versatile and more trouble-free ways of gasifying biomass and getting a syngas that gets as close to the optimum ratio of CO to H_2 as possible. In this, researchers now have achieved almost complete success.

A fully proven technology

While the fermentation of biomass to produce cellulosic ethanol is still in its infancy, the technology for gasifying biomass is fully proven for almost all types of biomass and is ready for commercial application. The earliest gasifiers were fixed bed gasifiers designed for coal. These consisted, basically, of the single vertical chamber with a combustion area just above the grate at the bottom, a feeding port for introducing the coal halfway or more up the chamber and a gas outlet at the top.

Later designs emerged out of the attempt to remedy the deficiencies of earlier ones and adapt them for gasifying biomass. In the 1980s, two scientists at the Institute of Gas Technology in Chicago, Suresh Babu and Michael Onischak, took an important step forward when they designed a pressurised, single chamber, fluidised bed gasifier that could handle a wide variety of biomass.[5] This gasifier, named RENUGAS, quickly became the standard biomass gasifier for most subsequent research.

Researchers soon found that gasifying biomass posed a different set of problems from those they had encountered in gasifying coal. Most of the problems arose from differences in its weight, its moisture content and the amount of inorganic matter that it contained.

Light and fibrous biomass, such as rice straw and husk, and the finely shredded bagasse currently fed into boilers in modern sugar mills, posed problems when fed into fluidised bed gasifiers because the air or oxygen blast tended to blow some of the fibre into the feed intakes and clog them. Pelletisation provided a solution to the problem but added to the capital cost of the process. The moisture content of biomass also tended to vary substantially from season to season. This had to be constantly monitored and offset by installing pre-drying facilities or introducing steam. However, the vast amount of heat generated by gasification made these adjustments relatively simple.

The third problem, a high silica content, required a different solution: since most of the first generation gasifiers worked at temperatures of 1,000°C or more, and silica melts at around 800°C, attempting to gasify some crop residues such as rice straw and husk tended to melt the silica in the straw while it was still in the feeding chute, and clog it, thereby stopping the reaction. Lowering the gasification temperature to 800°C or less prevented the silica from melting but increased the proportion of methane and heavier gases in the raw gas (called producer gas) that had to be 'cracked' into CO and hydrogen. This added to the cost of the process.

In 2002 a benchmark survey of the state of gasifier technology by the US National Energy Technical Laboratory (NETL) in 2002, listed four basic gasification systems then in use. These were fixed bed, updraft and down draft gasifiers and bubbling and circulating fluidised bed gasifiers. After studying 15 applications of these technologies it concluded that the bubbling fluidised bed gasifier had proved the most successful design for converting biomass into synthesis gas.

The survey had confirmed that to obtain the highest yield of synthesis gas from the biomass, gasification had to be done at 1,200° to 1,300°C, because at these temperatures all heavier hydrocarbon gases were broken down into CO and H_2, leaving little or no methane or heavier gases to be purged or cracked afresh. Synthesis gas could thus be obtained from a single pass of the biomass through the gasifier.[6]

The majority of these technologies had been developed to gasify MSW (garbage) whose safe disposal had become a serious problem in the industrialised world, but several had been developed specifically to gasify other forms of biomass such as bagasse, corn stover, wood chips and shavings, and black liquor, a toxic effluent of the paper industry.[7]

Developments in the last decade

Research on the gasification of biomass got another fillip after oil prices rose sharply once more in 2004. So rapid was the progress that in less than a decade since the NETL study was out of date. The focus of research shifted to increasing the versatility of gasifiers in order to minimise the number of stages through which the biomass-to-transport fuel process had to go. Municipal solid waste posed the most serious challenge because its composition changed from season to season; was different in rich to poor countries, urban and rural communities, and large and small towns.

Plasma gasification

Perhaps the most promising recent advance in gasification has been the use of plasma torches to gasify solid wastes.[8] Plasma is an ionised gas that is formed when a strong charge of electricity is passed through a normal gas. In nature, this happens every time lightning rips through the atmosphere. Ionised gases differ from ordinary gases in one important respect—they respond to magnetic charges and can therefore be shaped into beams or torches by an appropriate placement of strong magnets. Plasma torches were developed by NASA originally as a part of its space programme, to simulate the heat that would develop when a spacecraft re-entered the earth's atmosphere. They were first used commercially to develop blade coatings for the fans of gas turbines, which have to withstand very high temperatures. In the 1980s, plasma torches began to be used to melt metals. The focus of research shifted to MSWs in the 1990s.

Over an ensuing period of 20 years, the Westinghouse Electric Corporation carried out extensive research to develop plasma torches for commercial use. The result has been a plasma gasification process that is gaining acceptance rapidly among municipal authorities as a safe and economical way of disposing of MSW.

The Westinghouse plasma process can gasify any biomass, from sewage sludge to solid wastes, at around 3,000°C and yield raw producer gas at a temperature of 1,000°C or more. At this temperature virtually, the only gases it contains apart from CO_2 are CO and hydrogen. Plasma gasification has many advantages over other processes: First, it requires no separation of the inorganic matter. On the contrary, the very high gasification temperature converts all inorganic matter into a vitreous residue that can be used as a building material.

Second, it needs no sorting, pre-drying, or pelletisation—only a shredding or pulverisation of the biomass. The higher is its moisture content the greater is the proportion of hydrogen to CO in the synthesis gas.

Third, it destroys all the dangerous toxins, which are released when solid waste is buried in landfills or incinerated, by breaking them down into non-toxic compounds and gases. In the first plasma gasification plant, a 24 tonnes-a-day plant built at Yoshii in Japan, it was found that the plant dioxins released by the process amounted to barely 1 per cent of the dioxins released by the most sophisticated incineration of the waste.

Not surprisingly, therefore, the first plants were designed to deal with hazardous wastes, including nuclear waste. As of the end of 2009, there were two 68 tonnes-a-day gasification plants being built in Nagpur and Pune in India and one 144 tonnes-a-day plant in Istanbul, Turkey.

But a large number of other municipalities all over the world had realised that this was an excellent way to dispose of not just hazardous but all forms of solid wastes. Alter NRG, a Canadian company that had in the meantime bought the Westinghouse Plasma

Corporation, had 40 project proposals for gasifying an average of 1,000 tonnes of MSW per day from North American cities. Half of these were designed to generate electricity; the remainder were to produce synthesis gas for liquid fuels and other industrial uses.[9] The largest of these was a 2,000 tonnes/day waste disposal plant in St. Lucie County, Florida, USA.[10]

Regrettably, the onset of global recession in 2008 and the crash of oil prices that followed put all but a few of these projects on hold. But research into plasma gasification continued. In the past decade it has gone through four stages of development. In the last, computer-controlled sensors constantly fine-tune the heat in all parts of the gasification chamber to small changes in the composition of the waste in order to ensure an unvarying composition of the producer gas. They also adjust the composition of the gas to meet the requirements of the transport fuel that is being produced. In addition to yielding synthesis gas, plasma gasification produces 50 per cent more heat than incineration.[11] After deducting its in-house use of power, it turns 80 per cent of the energy in the solid waste into usable energy whether as heat or producer gas.[12]

Black liquor gasification

A second, fully proven gasification process is the Swedish Firm Chemrec's Black Liquor Gasification (BLG) process. It was developed by a consortium of five Swedish companies for producing transport fuels from black liquor as part of the European Union's ALTENER project for producing alternate transport fuels from biomass. Black liquor is a carbon rich but toxic amalgam of alkalis and chemicals obtained when wood pulp is bleached for the manufacture of paper. In modern paper mills, the carbon is removed from the bleaching liquor in a recovery boiler to yield what is known as 'green liquor'. This is further processed to take out other chemicals used in the bleaching process to restore the bleaching liquid to its original state—known as white liquor—for reuse.

In modern paper mills, the 'black' portion of the liquor is fed, in the form of air-dried briquettes, into the boilers to provide heat. The Chemrec-led research team developed a proprietary gasifier design process at New Bern in Sweden in a plant that went on stream in 1997 and, by the time it was shut down in 2008, had been in operation for 47,000 hours. The team used the pilot plant to test every element of the design of the gasifier, the most important being the longevity of the refractory bricks that were needed to line the gasifier, much as they are used in a blast furnace at a steel mill. The New Bern gasifier was shut down only after it had been operated continuously for two years, with only 5 per cent of 'downtime' for repairs and maintenance.

The quest for versatility—ZSW's AER gasifier

Other gasifiers have developed that target mainly difficult crop residues, such as rice and wheat straw, and rice husk. Here too the goal has been to progressively increase the versatility of the gasifier. At the Institute for Solar Energy and Hydrogen Research (ZSW) in Stuttgart, researchers have developed an absorption enhanced reforming (AER) fluidised bed gasifier that is capable of yielding the precise ratio of hydrogen to CO that is required to produce the desired transport fuel, be it methane, methanol, DME, gasoline, diesel or aviation spirit, in a single 'pass' of the biomass through the reactor.

The AER process has been developed from the design of an existing dual chamber fluidised bed (DFB) biomass gasifier that, in 2010, had been producing electricity from a wide variety of biomass inputs including sewage sludge and MSW, for the town of Güssing in Austria for the previous eight years. The Güssing plant has been gasifying 2.2 tonnes of air-dried biomass per hour to produce 2.5 MW of electricity and 5.5 MW of thermal energy and has achieved an overall energy efficiency of 80 per cent.

ZSW's research team headed by Dr Michael Specht, one of the pioneers of biomass gasification, has adapted the Güssing gasifier to produce liquid fuels by raising or lowering the temperature of the reaction and adding steam and lime (calcium oxide) to the fluidising medium.[13] The purpose of these adjustments is to increase the proportion of hydrogen to CO, which is usually too low in the raw gas obtained from biomass and to remove the surplus CO_2 in it. The extra hydrogen is obtained by lowering the gasification temperature to obtain more CO and then combining the CO with steam in what is known as a water gas shift reaction to break it down into hydrogen and carbon dioxide. This extra CO_2 and most of the original CO_2 in the raw gas is removed by the lime which combines with it to yield chalk.

The AER gasifier is even more energy-efficient than the DFB gasifier at Güssing because the chemical reaction by which lime combines with the CO_2 to form chalk is also exothermic. Overall, therefore, well over 80 per cent of the energy contained in the biomass gets converted into transport fuels. Specht and his colleagues presented the results of their trials with the AER gasifier at a conference on Polygeneration Strategies at Vienna in September 2009.[14]

Notes and references

[1] The US-based Sierra Club has been a tireless campaigner against coal fired or coal gasification-based power projects. Between October 2005, when it helped Manistee, a small lake Michigan community to stop the Texas-based Tondu Corporation from constructing a large coal-based power plant in its midst, and April 2012, it has prevented coal-based plants from coming up in 61 cities and communities across the United States. See also James Hansen's testimony to the US Senate in 2008 and chapter of his book *Storms of my Grandchildren*. See http://www.sierraclub.org/environmentallaw/coal/victories.aspx-#illinoisindeck (accessed on 10 July 2017).

[2] This is described in greater detail in Chapter 6.

[3] George A. Olah, Alan Goeppert and G.K. Suryaprakash, *Beyond Oil and Gas: The Methanol Economy* (Weinheim: Wiley-VCH. Verlag GmbH & Co. KGA, 2006).

4   International Energy Agency, 'Task 33 Presentation to EXCo 59 in Golden, Colorado 25–27 April 2007: Summary of Discussion at the Task 33 Workshop on Situation Analysis and Role of Biomass Gasification (BMG) Technologies in Future Energy Needs', available at http://www.ieatask33.org/app/webroot/files/file/publications/ IEA_BRU_11-07.pdf (accessed on 2 September 2012).

5   Details of the RENUGAS gasifier may be found at http://www.ficfb. at/aer.htm (accessed on 10 July 2017).

6   The NETL study concluded: 'Ideally, for fuels, chemicals and hydrogen applications, it is beneficial to operate at high temperatures. At temperatures greater than 1200–1300°C, little or no methane, higher hydrocarbons or tar is formed, and H_2 and CO production is maximized without requiring a further conversion step'.

7   Jared P. Ciferno and John J. Marano, 'Benchmarking Biomass Gasification Technologies for Fuels, Chemicals and Hydrogen Production', prepared for US Department of Energy (National Energy Technology Laboratory), available at http://www.netl.doe. gov/technologies/coalpower/gasification/pubs/pdf/BMassGasFinal. pdf (accessed on 10 July 2017). A summary of the survey results is reproduced as follows:

Table 6 lists gasification operating conditions for fifteen technologies for which sufficient data were available. Of the technologies listed, seven are bubbling fluidized bed (BFB) gasifiers, six are circulating fluidized bed (CFB) gasifiers and two are fixed-bed (FB) updraft gasifiers. The majority of the processes listed have been tested with a variety of biomass feedstocks. However, results have only been reported for a few different feedstocks, and it is believed that many of the feedstocks reported have only been tested in small-scale bench units. The primary feedstocks, for which product syngas composition data were available are identified in Table 6. These were typically wood, pulp sludge, MSW, RDF and corn stover. The feed rate ranged from 136 to 7,575 kg/hr (300–16,665 lb/hr); pressure from 1 to 33 bar (14.7–480 psi); and average reactor temperature from 725 to 1400°C (1337–2550°F). Table 7 summarizes the ranges of conditions tested for the various biomass gasifier classifications: BFB (directly heated), CFB (directly heated), fixed bed, indirectly-heated CFB (BCL/FERCO), and indirectly heated BFB (MTCI). For comparison, Table 7 also includes typical operating conditions for the commercial Shell entrained-flow gasifier. The Shell coal gasification process has been demonstrated at a throughput that is an order of magnitude greater than normally encountered with biomass. The availability of large quantities of coal at centralized locations enables coal gasification facilities to take advantage of economies of scale. Operating biomass gasifiers at or above atmospheric pressure has both benefits and drawbacks depending upon the intended application for the syngas. Pressurized gasifiers are complex, costly and have a higher capital cost, both for the gasifier and associated feed system. On the other hand, the gas supplied to a combustion turbine or conversion process is at pressure, avoiding the need for costly gas compression. Exit temperatures vary considerably reflecting gas clean up and heat recovery systems. Some investigators have only reported temperatures downstream of this equipment. Sources of oxygen used in biomass gasification are air, pure oxygen and steam, or some combination of these. Air is the most widely used oxidant, avoiding the requirement for oxygen production, and was used in over 70% of the gasifiers that have been tested. However, the use of air results in a low heating value gas, 4 to 6 MJ/m³ (107–161 Btu/ft³), only suitable for boiler and engine operation. The use of oxygen produces a medium heating value gas, 10 to 15 MJ/m³ (268–403 Btu/ft³), suitable for combustion turbine applications or for conversion to gasoline or methanol [9]. The BCL/FERCO and MTCI gasifiers produce the highest heating value syngas with 16.7 to 18 MJ/m³.

8   Plasma, referred to as the 'fourth state of matter', is a very high temperature, highly ionised (electrically charged) gas capable of conducting electrical current. Examples of plasma in nature include lightning and gas at the surface of the sun. Plasma technology has a long history of development and has evolved into a valuable tool for engineers and scientists who need to use very high temperatures for new process applications. Man-made plasma is formed by passing an electrical discharge through a gas such as air or oxygen (O_2). The interaction of the gas with the electric arc dissociates the gas into electrons and ions, and causes its temperature to rise, often to exceed 6,000°C. This is nearly as hot as the sun's surface.

9   Alter NRG Corporation and Westinghouse Plasma Corporation, 'The Plasma Gasification Process', presentation by Alter NRG at the Calgary Regional Partnership Waste Summit (9 April 2008).

10   The onset of global recession made raising the capital for such a large plant difficult. This led first to a reduction in its size and then its

abandonment. Many other renewable energy projects suffered the same fate. This is described in greater detail in the appropriate sections of this and the next chapter.

11   Erwin Altmann and Paul Kellett, *Thermal Municipal Solid Waste Gasification: Status Report* (Bandon, County Cork, Ireland: Irish Energy Centre Renewable Energy Information Office).

12   US Department of Energy, The Energy Lab, 'Where Energy Challenges Converge and Energy Solutions Emerge', Gasifipedia, available at http://www.netl.doe.gov/technologies/coalpower/gasification/gasifipedia/4-gasifiers/4-1-4-1a_westinghouse.html (accessed on 10 July 2017).

13   Usually olivine, a commonly found magnesium iron silicate.

14   N. Poboß, A. Schuster, G. Scheffknecht, 'Experimental Results on Hydrogen Production from Biomass Applying the Absorption Enhanced Reforming (AER) Process in a 20 kWth Dual Fluidized Bed', poster presentation, ICPS International Conference on Polygeneration Strategies 09 Vienna (2009).

Chapter Seven
Limitless Potential of Biomass

Contrary to a widespread belief, there is more than enough waste biomass in the world, in the form of crop residues and municipal solid waste to meet the entire world demand for transport fuels, for a long time to come.

The three processes for biomass gasification described above are among the most recent of at least a round dozen that are in commercial or small scale use in various parts of the world. It is, therefore, safe to claim that gasification of virtually any form of biomass is now a technologically proven process. But three more conditions still have to be met for it to make a significant dent in CO_2 emissions. First, there must be enough biomass to allow a substantial replacement of fossil fuels. Second, it must not disrupt the way people live and work, and relate to each other. Third and most important, it must be profitable to the investor and not only to society. Only then will industry invest in it without the benefit of subsidies.

Is there enough biomass?

The knee-jerk response to the first question is usually a 'no'. In *The Revenge of Gaia*, James Lovelock dismissed biomass as a possible alternative to fossil fuels because, he believed, 'there is not enough of it'. James Hansen is also of the same view for in *Storms of My Grandchildren*, he has conceded, albeit reluctantly, that there is no alternative to oil and gas as a transport fuel in the medium-term future. He has, therefore, urged governments to concentrate on phasing out the use of coal by 2030.

Others are dismissive to the point where they do not even mention it. When the German government decided, in the wake of the Fukushima disaster, to phase out nuclear power altogether, its spokesmen spoke at length of the potential of wind and solar power to replace the plants that would be closed down, but did not mention ZSR's work on biomass gasification. None of the major reports on climate change of the last 15 years, such as the ExxonMobil energy projections, the Stern report, the IPCC's Fourth and Fifth Assessment Reports, and Tony Blair's 2009 report on 'Technology for a Low Carbon Future', even mentions 'methanol'. The ExxonMobil report only mentions cellulosic ethanol. The IEA's Biofuels II Report concludes that 'second-generation biofuels', a term that covers cellulosic ethanol and 'Fischer–Tropsch fuels', will be available commercially only after 2030. China is the only country with an active methanol production programme for use as a transport fuel. But it is producing it from coal and may therefore speed up the world's rush to disaster.

Ordinary people discount biomass because they believe that most of it already has important alternative uses. For instance, commercial timber does not form much more than half of the total biomass in a tree. But the remainder of the wood is used to produce charcoal, wood pulp (for paper), chipboard or burned as firewood. Similarly, cotton bolls are separated from the seed to process further into yarn, but the seed is crushed to yield cottonseed oil, and the residue is fed to cattle as an important nutritional supplement. This is true of other oilseeds as well. In many poor countries, a part of the rice and wheat straw left behind after harvesting is kept in reserve to feed cattle in case a flood or a drought destroys their normal pasturage. Bagasse, the residue from sugar cane, is fed into sugar mill boilers to produce the heat needed for distilling the cane juice. Lentil stalks are routinely fed to cattle. Even black liquor, a toxic but carbon-rich effluent from the paper industry is dried into briquettes and fed back into the paper mills' boilers to provide the electricity and process heat needed by the production process. But, as is elaborated below, even the best of these uses adds only a small amount to value of the primary product, be it food grains, cash crops, sugar or paper. Gasification permits these residues to be harnessed to the most profitable and relentlessly growing sector of the global economy, the demand for transport fuels.

In addition, there are many other forms of biomass that are of no use whatever. For instance, rice husk is simply burnt. Rice straw has almost no nutritional value because it contains very little nitrogen and a very large amount of silica. Wheat straw is a little more edible because it has less silica but also has little nutritional value. In the main rice-growing countries, tractors have progressively replaced draft cattle and reduced the need for cattle feed. This has made rice and wheat straw redundant. In the paddy fields of India and Pakistan, the stubble is regularly burned in the fields after the harvest to clear the way for winter wheat. Half of the sugar cane crop consists of waste that cannot be burned in modern bagasse burning boilers use, but can be gasified. Farmers regularly squabble with mill managers over the responsibility for disposing of it. Eventually it rots or is burned, partly at the sugar mills and partly in the fields. According to an estimate made in the late 1990s, farmers in India were burning at least 100 million tonnes of crop residues in their fields every year. This figure has almost certainly doubled since then.[1] In 2001, the EU alone had 700 million tonnes of agricultural wastes.[2] These probably exceed a billion tonnes now.

Municipal solid waste

Lastly, there is an abundance of biomass that not only has no use, but poses a danger to human health. This is MSW, whose disposal has become the single most serious problem that cities all over the world face. Depending upon their affluence and degree of urbanisation, human beings generate between 1 and 2 kg of garbage every day. Rising incomes change both the amount and its composition. In low income countries, most of the garbage consists of kitchen wastes. In industrialised ones, most of it consists of paper, plastics, discarded consumer durables, furniture and clothing, but also batteries, medical wastes and toxic rare earths used in computers. This has made it more and more difficult to dispose of solid wastes in safe ways. The prevailing practice is to bury it in landfills or burn it. New York City, where space is at a premium, incinerates 38 per cent of the trash it collects and exports the rest to upstate landfills.[3] As recently as 2004, Tokyo and Hong Kong, which generated 1.13 and 2.2 kg of trash per person per day, respectively, were burying a quarter to three-fifths of their garbage in landfills.[4] The figures for smaller cities across the developed world are worse because they believe that they have land to spare. New York, for instance, does both. In poorer countries, municipalities increasingly do not have the resources to collect the ever-mounting piles of garbage.[5] So while what is being collected is still being buried in landfills, a sizable portion is left to rot in the sun.

Burying waste in landfills is extremely unsafe because several toxic compounds leak out of decaying organic matter and seep into underground water reservoirs. But this option is rapidly being closed by the sheer non-availability of land. In the 1970s and 1980s, there was a sharp rise in the number of incineration plants all over the world, because of the spike in oil prices, but environmental concerns, enshrined in new pollution standards soon put a brake on them. Cities in high income countries now require households to sort the garbage before they collect it but are seldom able to reuse more than a third of it. The balance therefore still goes into incinerators or landfills.

In the low income countries, garbage is dumped in insanitary landfills or left to litter the sidewalks and rot in the sun. In spite of this, the municipal authorities in New Delhi, India, collect more than 10,000 tonnes of MSW every day. This is sufficient to cover a football field more than four metres deep in muck.[6] In the years ahead, the quantum of urban solid waste will continue to grow, and its growth too will be exponential.

If it is turned into syngas instead of being burned, it will not only solve all the problems of disposal but will be able to replace a large part of the transport fuels being derived from fossil fuels today. Today, only the unprecedented depth of the global recession stands in the way.

Crop residues

The other vast reservoir of immediately available biomass is crop residues. In 2017–2018, the global output of cereals is expected to be 2.611 billion tonnes.[7] Since a tonne of grain leaves behind, on an average, a tonne and a half of straw and other agri-wastes, the crop residues from cereals alone will come close to 4 billion tonnes.[8] The residues from six principal oil-producing crops amount, in most years, to an additional 1.5 billion tonnes. In 2011, the top 10 sugar cane-producing countries in the world grew 1.794 billion tonnes of cane which yielded 500 million metric tonnes of bagasse and over 1.5 billion tonnes of sugar cane leaf and other waste.[9] The paper industry produces approximately 175 million tonnes of black liquor cakes that consist almost entirely of carbon leached out of wood pulp before making paper.[10] Overall, therefore, the principal agricultural crops grown around the world generate close to 8 billion tonnes of crop residues a year.

If all of these crop residues were to be gasified using oxygen in place of air (process V-2 in Table 7.1 below), they would yield more than 4 billion tonnes of methanol, equivalent in kinetic energy to 3 billion tonnes of gasoline. This is only a shade less than the entire global consumption of transport fuels. These are only the currently most commercialised crops, whose residues are already being burned in boilers or in farmers' fields. Lovelock's dismissal of biomass as an alternative to fossil fuels was, therefore, premature, to say the least.[11]

The shift from fossil to biofuels would, therefore, virtually eliminate urban litter and the diseases it spreads, and sharply reduce vehicular smog and the incidence of the diseases ranging from the common cold and bronchitis to asthma and cancer. The conversion of crop residues into transport fuels will also curtail the release of methane from the decay of agricultural wastes.

Profitability—Will industry adopt biomass gasification?

The profitability of investment in biomass gasification depends upon the capital investment required; the intended transport fuel; the sophistication of the process employed and the potential for co-producing the transport fuels with existing products. The last is greatest in agro-industries such as sugar and paper because these already have the power and transport infrastructure, the managerial resources and, most importantly, the biomass collection systems that are needed to provide an assured supply of feedstock to a continuous process biofuel plant.

Since the two most widely used transport fuels are gasoline and diesel, it is not surprising that most of the economic appraisals made so far have been of the cost of producing methanol which is a substitute for gasoline and DME which is a substitute for high-speed diesel. However, these are no longer the only fuels that are slated for commercial production from biomass. Several projects are on the anvil for the production of aviation turbine fuel diesel, Naphtha and other 'Fischer–Tropsch' fuels.

One of the earliest estimates of the cost of producing transport fuels from biomass was presented by Michael Specht, A. Bandi, F. Baumgart, C.N. Murray and J. Gretz, scientists belonging to three European institutions—the ZSW in Stuttgart, the Environment Institute's Joint Research Centre in Ispra (Italy) and the International Association for Hydrogen Safety in Hamburg

(Germany). This study, which became a benchmark for future research, compared the yield of methanol from a tonne of air-dried biomass (containing 10 per cent moisture) with the capital and operating costs of producing it using three technological production options: (a) when oxygen for the gasification process is extracted physically from the air; (b) when the oxygen is obtained from the electrolysis of water and the by-product hydrogen is also fed into the gasifier; and (c) when additional hydrogen is added, possibly by investing in a larger electrolysis plant, to minimise the output of CO_2. For the sake of convenience, these are labelled V-1, V-2 and V-3, respectively.

Specht and his research team found that in their 50 tonnes-a-day-of-biomass pilot plant, V-1 yielded 0.3 tonnes of methanol per tonne of air-dried biomass, V-2 yielded 0.6 tonnes and V-3 yielded 1.1 tonnes.[12] Their results conformed closely to estimates that had been made six years earlier by the US Department of Energy based upon experiments in Hawaii which had shown that oxygen-blown gasification (V-2) converts a short tonne of biomass into 186 gallons (572 kg) of methanol.[13]

Table 7.1 presents their economic assessment of the three processes. It shows that the choice between the three versions will depend not upon the capital cost of the plant, where the difference between the cheapest and most expensive is only 50 per cent, but on the cost and availability of power. Specht et al. based their

Table 7.1: Capital costs power requirements and yield of methanol under various assumptions in 1997

Variants	Methanol/ Tonne Biomass	El. Consp. (MW)	Total Investment (m. $)	Cost/Tonne Methanol ($)
V-1	0.3	1.1	16	685
V-2	0.6	5.9	18.3	511
V-3	1.1	14.7	24.8	444

Source: Author's compilation.

calculations upon the capital cost of the required equipment in 1997 and an electricity cost of 3 US cents (0.05 pfennigs) per unit and found that while V-3 quadrupled the methanol output, it increased the consumption of electricity by almost 14 times. The median option (V-2) doubled output but increased electricity consumption by 5.3 times.

In spite of this, V-3 turned out to be the cheapest option for producing methanol.

Specht's calculations showed that using V-2, the price of 2.2 litres of methanol (which had the same energy content but far greater kinetic force than a litre of gasoline) came to 88 US cents, or about 0.9 euros. While this was considerably higher than the 26–30 cents a litre that prevailed in most American states, it would have been far cheaper than the $1.11 that prevailed in France and even higher prices in several other European countries, at the time!

Seven years later, in a 2005 review article assessing the findings of various research groups, P. Galindo Cifre and O. Badr of the Engineering Study Group of Cranfield University in the United Kingdom estimated that methanol could be produced from biomass at 300–400 euros a tonne. This was a good deal higher than its cost of production from natural gas, which was then about 200 euro a tonne,[14] but by then the price of gasoline at the pumps had risen to approximately 1.20 euro per litre or about 1,550 euro per tonne. They too concluded, therefore, that producing transport fuels from biomass would be profitable if European governments did not levy some of the taxes they had saddled upon gasoline.

By the beginning of 2014, the economic scales had tilted even more heavily in favour of methanol and DME because, while the cost of investment (estimated from the chemical plants cost index) had gone up since the beginning of the century by 60 per cent, and the cost of electricity had trebled, the price of crude oil had gone up by five times to over $100 per barrel.

Co-production I: The example of black liquor gasification

In December 2003, as part of an EU ALTENER project designed to suggest ways of shifting 12 per cent of transport fuel consumption to biofuels by 2010, five firms in Sweden undertook a two-year research project to study the technical and economic feasibility of converting black liquor into methanol for use as a blending fuel with gasoline. The research was carried out with simulation models developed from data gathered in a 50 tonnes/day pilot plant using the Chemrec gasifier developed in the late 1990s that was described in the previous chapter. The purpose of the simulation was to determine the economics of modifying a 2,000 tonnes/day paper pulp mill to produce methanol as a by-product.[15] The pilot plant studies showed that gasifying the black liquor instead of drying it into cakes and burning it in the existing boilers generated all the process heat that was needed to run the paper mill and enough surplus heat and power to run the methanol reformation plant as well. The pilot plant showed that the methanol plant attached to a paper mill would yield 1.4 to 1.6 tonnes of methanol for every tonne of dried black liquor. If an electrolysis plant were added to produce both the oxygen that the reaction needed and the hydrogen required for the complete conversion of the carbon in the gas to methanol, the yield would go up to 2.2 tonnes of methanol for every tonne of black liquor.

As the lead company in the research group, Chemrec submitted its BLG process to the EU Commission for approval as part of the ALTENER project for replacing 12 per cent of oil- and gas-based transport fuels with biofuels. The EU cleared the process for commercial use after concluding that there were no technical obstacles left to be overcome. By 2006, two demonstration plants had been set up in Växjö and Piteå in Sweden. Not surprisingly, therefore, projects involving biomass gasification on a commercial scale began to appear on the industrial horizon in several countries. Indeed, had the global recession not intervened to set the calendar back Europe, the United States and many other countries both within and outside the OECD would have been producing sizable amounts of methanol and DME to blend with gasoline and replace diesel today.

The ALTENER study also confirmed that even without adding additional hydrogen (process V-2 above) a 2,000 tonnes/day pulp mill would be able to produce 2.2 litres of methanol, equivalent in energy content to a litre of gasoline, for 28.7 centimes, and DME equivalent to a litre of diesel for 31.8 centimes and would recover its investment in the gasification and methanol/DME reformation plants in four years.[16] Price sensitivity studies made as a part of the estimation of costs and returns showed that the cost of production of both fuels would rise by 6.6 per cent for a 30 per cent rise in the capital cost and by 8.2 per cent for a similar rise in the cost of purchased electricity. The 60 per cent rise in capital costs and trebling of electricity prices around the globe in the past decade is likely to have pushed these estimates to between 50 and 60 centimes a litre. This is not only less than a third of the pump price of gasoline in any European country, but below current pump prices in the United States as well.

In 2007, Chemrec entered into an agreement with the US paper industry giant the NewPage Corporation to build a 600 tonnes/day black liquor-based methanol plant at the latter's Escanaba Pulp and Paper Mill in Michigan, USA, but the world went into recession the very next year and the demand for methanol both by the petrochemicals and transport fuels industries vanished. As a result, the project was shelved in July 2009.[17]

In 2010, the Arlanda Airport authority in Sweden completed a technical and financial feasibility study to build two plants near Stockholm and Arlanda that would produce 50,000 tonnes of ATF and a synthetic crude oil that would be further refined into a variety of transport fuels. In all, the two plants were to produce up to 180,000 tonnes of jet fuel and other Fischer–Tropsch products. The study showed that the jet fuel plant would not be profitable without a subsidy at the then reigning price of oil at $67 a barrel, but would become so if its price rose (as it later did) above $100. However, the synthetic oil plant would be competitive even at the 2010 price of oil.[18]

In 2010, Chemrec entered into an agreement with Domsjö Fabriker, originally a pulp mill that had become a pioneer in the production of specialty cellulosic products, to set up a third plant at Örnsköldsvik in northern Sweden to produce 140,000 tonnes of bio-methanol or 100,000 tonnes of DME or methanol for use as a surface transport fuel.

But the Invisible Hand, or is it fist, of the Market Economy struck again. Oil prices collapsed once more, for the third time since 1985, at the end of 2014. As a result neither project has come on stream. The Arlanda Airport authority postponed the construction of the two plants near Stockholm, and in 2012 Domsjö Fabriker withdrew from the Örnsköldsvik project after it was taken over by an Indian conglomerate the Aditya Birla Group which was primarily interested in Domsjö's vast expertise in the production of specialty cellulose products.

DME as substitute for diesel—The volvo trials

Recognising that getting its biofuels accepted by the users is as important as developing them, in 2009 Volvo and Chemrec teamed up with two road haulage companies in Sweden PostNord and DHL and ran five DME-fuelled heavy trucks each for over two years to assess the performance of DME as a substitute for diesel. By May 2012, the trucks had covered an average of 80,000 km each, hauling 60 tonnes of paper per load and reported no significant problems. Greger Hallgren, one of the drivers, reported an almost complete absence of soot in the exhaust pipe after 70,000 km of running and a considerably greater power delivery from the same engine. When interviewed by Chemrec for its in-house journal, he said, 'DME is' a more powerful fuel than diesel'. He added, 'This engine of 440 HP is much stronger in the uphill slopes than a 500 HP diesel. It is also more quiet!'

The environmental affairs manager of PostNord, Henrik Boding, also reported, 'Our drivers are very pleased'.

> They report that it is at least as easy to run on Bio-DME as it is on conventional diesel fuel. This is an entirely new technology but we have nevertheless experienced very few technical problems and, what is more, the trucks run much more quietly with Bio-DME in the tank.[19]

A well-to-wheel emissions study, sponsored by Volvo that took into account the CO_2 and other GHGs generated by the energy needed to produce the alternative fuel and the GG impact of the land use changes that accompanied the adoption of various biofuels, found that substitution of diesel by DME and gasoline with methanol-reduced carbon emissions by 95 to 97 per cent in the case of DME and 89 to 92 per cent in the case of methanol. The study also found that the highest GHG emissions came from the two 'first generation' biofuels that the entire world is striving to produce—ethanol and biodiesel.[20]

Co-production II: The commercialisation of plasma gasification

The other gasification technology that is rapidly coming into commercial use is plasma gasification. This too owes a great deal of its attraction to the fact that it produces not one but three products, two tangible and one intangible. The tangible ones are energy and land. The intangible is public health.

As was described earlier, the sheer versatility of this technology, its ability to gasify any biomass and its high ratio of hydrogen to other gases in raw producer gas that has to be converted into synthesis gas have drawn worldwide interest. Plasma gasification is exceptionally well suited for gasifying MSW as its very high flame temperature (>3,000°C) makes pre-sorting the solid waste to remove tins, bottles and other inorganic matter unnecessary. When it is carried out with oxygen, it produces only 7 per cent of the flue gas obtained from combustion. The reaction takes place, moreover, at such high temperatures—1,000°C to 3,000°C—that all the dioxins and furans get broken down into their basic elements and completely lose their toxicity. In a 24 tonnes/day plasma gasification plant that has been running for more than a decade at Yoshii, in Japan, the release of dioxins has been found to be less than '1 per cent' of what is released by the best available incineration plants. According to the US-based Recovered Energy, a turnkey engineering company specialising in renewable energy projects, there are 200 MSW gasification plants under construction or in operation in the world, of which half use the revolutionary new technology called plasma gasification.

Therefore, it is not surprising that its use has widened from the destruction of hazardous wastes (India, Turkey and Japan) to the safe disposal of MSWs. An agreement for the first large project designed to do this in St. Lucie County Florida (2,000 tonnes, 160 MW) was followed by one for a 2,500 tonnes/day plant, designed to produce 138 MW of power, at New Orleans in November 2006, and one for a 1,000 tonnes/day, 35 MW plant at Tallahassee, Florida, in early 2007. In Tallahassee, the city authorities drove a hard bargain with the investors Green Power Systems and contracted to pay it only 5.98 cents a unit against the national average power price of 9.1 cents. Green Power Systems did not publicise details of their operating costs and revenues but an economic appraisal based upon prevailing norms for investment suggested that the venture would yield a return of more than 4 per cent a year to equity shareholders. If the price it received had been 9.1 cents, the return on equity would have been in excess of 13 per cent.[21]

According to the US-based *Recovered Energy*, a turnkey engineering company specialising in renewable energy projects, in 2013 there were 200 MSW gasification plants under construction or in operation in the world, of which half used the revolutionary new technology called plasma gasification.[22]

In 2014, plasma gasification was about to be deployed for producing transport fuels on a large commercial scale. Three years earlier, Solena, a US-based technology firm established at the turn of the century, announced a collaboration with British Airways (BA) to set up a waste-to-transport fuels project, aptly named GreenSky, that would gasify 1,300 tonnes a day of London's solid waste to produce more than 16 million gallons of ATF and 9 million gallons of diesel 'in addition' to generating up to 40 MW of power. To reassure investors that their money would be safe, British Airways signed an 11-year purchase agreement with Solena to purchase their aviation turbine fuel.

BA's project inspired Qantas to approach Solena for a similar project in Australia and Scandinavian Airlines System (SAS) to set up a similar waste to biofuels plants in Sweden, Norway and Denmark.[23] Lufthansa also did so a little later. By 2013 Solena had signed a letter of intent with seven American airlines to provide them with biofuels from an almost identical plant it proposed to set up in Santa Clara County in northern California.[24]

But the depth of the global recession made raising the finance difficult. Solena sold a large part of its business to another American company. When oil prices crashed once more, the company was unable to raise the final tranche of investment it needed. As of 2017, the Greensky project is on indefinite hold.

The strange neglect of crop residues

Considering that methanol, the first alternate fuel to coal and oil, was obtained from biomass, its subsequent neglect has been hard to understand. As was described earlier, two of the most readily available feedstocks are black liquor, from the paper industry, and bagasse and sugar cane waste from the sugar industry.

The obstacles that have prevented the commercialisation of black liquor gasification have been described in the previous chapter. But the neglect of bagasse, in particular, is hard to explain. Bagasse is a far easier material to gasify than either solid waste or black liquor. Oven dry, it has the same energy content as good firewood:[25] Shredded bagasse comes out of the mill with up to 45 per cent moisture but this soon dries out to around 25 per cent, an ideal amount for maximising the hydrogen in the raw producer gas that gasification yields. Bagasse has less than 0.1 per cent of both chlorine and sulphur, so the raw gas is easy to clean. It also has an ash content of only 2 per cent so it creates very little residue.

Unlike municipal waste and black liquor, which require specialised gasifiers, bagasse can be gasified in any of several existing designs of gasifier with only minor modifications to the bagasse intake system. The sugar industry can be an ideal partner for the co-production of transport fuels and olefins. In modern mills, sugar cane is crushed, shredded, mixed with water and driven through rollers again till nearly all of the sugar it contains is squeezed out. The remaining bagasse is dried and fed into the boiler. The sugar cane juice is heated, concentrated and allowed to cool to yield crystalline sugar.

In this process, the mill also obtains two tonnes of molasses for every five tonnes of sugar. This is fermented to yield ethanol, which is distilled again to yield potable alcohol for the liquor industry. The process heat required by a sugar mill for all of these operations consumes only a quarter of the heat generated by the boilers, so the balance is used to generate electricity. About a quarter of that electricity is consumed within the plant. The balance is sold to the grid or to individual buyers. Today, the co-generation of electricity has become normal in the sugar industry all over the world.

But bagasse does not only burn cleanly: it is exceptionally for gasification because it contains virtually no chlorine or sulphur. So the producer gas it yields requires very little cleaning to ensure that it does not poison the catalysts used in its conversion into transport fuels.[26] It is, therefore, something of an anomaly that while BLG has been subjected to more than 15 years of development and achieved full commercial viability, the gasification of bagasse on a commercial scale has not even begun.

The reason is not far to seek. The sugar industry has been among the first to explore co-production of renewable energy alongside its main business. But in Brazil the industry has chosen to produce ethanol. In India, Pakistan, Bangladesh, Mauritius and other cane-growing countries, it has chosen to generate power and sell it to the national grid. Till the end of the 1970s, sugar mills considered bagasse a nuisance and burned it in their yards after extracting the juice from it. Only after the two oil price shocks of the 1970s did mill owners begin to look for ways to use it as a fuel, in place of coal and gas, to meet their need for process heat. Modern boiler designs, specially tailored to cope with the properties of bagasse (such as its extreme lightness when shredded) improved the efficiency of combustion till mills not only no longer needed coal but began to generate large amounts of surplus heat. They used this to generate power.

One reason why co-producing methanol may not have entered sugar mill owners' minds could be that their conversion to bagasse-burning boilers is relatively recent. In India for instance, nearly all of the 210 sugar mills that were using it as boiler fuel by 2010 had installed their modern boilers after the turn of the century. Another could be the very strong association of biofuels with ethanol in the public's mind. But with the limitations of ethanol exposed a return to producing methanol, DME and other transport fuels via gasification is long overdue. The global availability of black liquor cakes is of the order of 175 million tonnes. The global availability of bagasse and sugar cane waste is over 2 billion tonnes.[27] In sum, this one industry alone is capable of meeting a quarter of the current global demand for transport fuels.[28]

Biochar

Black liquor and bagasse can be used as feedstocks for producing synthetic transport fuels immediately because they are readily available as by-products at paper and sugar mills. But the straw from cereal crops and corn stover, which make up the bulk of the developing countries' crop residues, can only be transported economically for very short distances, so it is usually burned to clear the fields for the next crop (posing an additional health and climate hazard) or stacked beside the field to be used as hay. That puts a limit on the size of any plant that processes them. This problem can be circumvented by first turning the crop residues into biochar and briquetting it, thus making its transport to the synthetic fuels plant easy and economical.[29]

Biochar is a residue of the pyrolysis or gasification of cereal crop residues, and makes up 22 to 40 per cent of the original mass of the feedstock, depending on whether it is derived from wood wastes or straw 49 to 72 per cent of it is carbon. It is therefore a rich feedstock for the production of transport fuels.[30] In the industrialised countries, converting crop residues into biochar has been advocated as a way of extracting carbon from the air and burying it cheaply, in a stable form, in the earth. But this is short-sighted thinking by people in a hurry.

In developing countries, ploughing carbon back into the earth has been seen as a way of restoring some of its nutrients and increasing the capacity of microorganisms in the soil to resist attacks by pollutants such as pesticides, heavy metals and hydrocarbons.[31] But since most of the carbon has been obtained from the air via photosynthesis, mixing it with the soil year after year will end by poisoning it. The alternative is to gasify it in simple, air-blown gasifiers, use the lean gas as a domestic fuel and briquette the biochar for sale as a high grade fuel that in turn can be used as a feedstock for making biofuels. The clean fuel gas, and the extra income this will generate, can transform farm families' lives, particularly those of their women.

Despite these benefits, the production of biochar has not taken off because there is no market for biochar briquettes.

Notes and references

1   Rajeev Jorapur and Anil K Rajvanshi, 'Sugar Cane Leaf-Bagasse Gasificatiers for Industrial Heating Applications' (Phaltan, Maharashtra: Nimbkar Agricultural Research Institute), published in *Biomass and Bioenergy* 13, no. 3 (1997): 141–146.

2   United Nations Environmental Programme, Division of Technology, Industry and Economics International Environmental Technology Centre, *Developing Integrated Solid Waste Management Plan Training Manual: Assessment of Current Waste Management System and Gaps Therein*, vol. 2 (Osaka/Shiga), 2, available at http://www.unep.or.jp/ietc/Publications/spc/ISWMPlan_Vol2.pdf (accessed on 10 July 2017).

3   Concerned Citizens of Cattaraugus County, Inc., 'New York City's Garbage Crisis' (25 July 2006), available at http://concernedcitizens.homestead.com/fkfacts.html (accessed on 10 July 2017).

4   George Tsz Ho Chow, Hinson Wai Wang Lau, Anson Tsz Shan Leung and Tommy Wing Chun Law, 'Waste Treatment in Tokyo And Hong Kong' (Japan East Asia Network of Exchange for Students and Youths Program 2009 [JENESYS], 2009), available at http://www.edb.gov.hk/FileManager/EN/Content_7720/jenesys2009_groupreport_4a_e.pdf (accessed on 2 September 2012).

5   In the Khajuria, Yamamoto Morioka's study cited below the MSW collected in Pakistan as estimated to be 0.45 kg per person per day against 0.8 kg for Indian cities. This almost certainly reflects a much lower level collection, not consumption.

6   The average English football field has an area of $7,035/m^2$.

7   See http://faostat.fao.org/site/567/DesktopDefault.aspx?PageID=567#ancor (accessed on 10 July 2017).

8   One tonne of rice generates 1.5 tonnes of paddy straw. It also generates 0.3 tonnes of rice husk. Wheat generates no husk.

9   See http://www.illovosugar.co.za/World_of_sugar/Sugar_Statistics/International.aspx and http://en.wikipedia.org/wiki/Sugarcane (accessed on 10 July 2017).

[10] T. Ekbom, N. Berglin and S. Logdeberg, 'High Efficient Motor Fuel Production Biomass via From Black Liquor Gasification', available at http://eri.ucr.edu/ISAFXVCD/ISAFXVAF/HEMPPBB.pdf (accessed on 14 august 2012).

[11] US Energy Information Administration, 'International Energy Statistics', available at http://www.eia.gov/cfapps/ipdbproject/IED Index3.cfm?tid=5&pid=66&aid=13 (accessed on 10 July 2017).

[12] M. Specht, A. Bandi, F. Baumgart, C.N. Murray and J. Gretz, 'Synthesis of Methanol from Biomass/CO_2 Resources' (22 August 2012). The calculations are based on the following assumptions. The feedstock for the gasifier is a wood input of 2 t/h with a water content of 10 per cent. The efficiencies are related to the lower heating value of methanol. All costs are given in Deutschmark (1 DM = US$0.56; August 1998). The costs are mainly determined by the energy and biomass input and the capital costs of the production plant. Operating and maintenance costs were estimated as 2.5 per cent per year of the investment costs. For all vectors, it was assumed that hydroelectricity is available at 0.05 DM/kWh. Capital costs are calculated on a real interest rate of 8 per cent and depreciation periods of usually 15 years. The plant capacity was calculated to produce about 25–50 tonnes of methanol per day. To evaluate investment costs and methanol production costs, a raw gas stream of 3,000 me/h is considered. The feedstock costs for natural waste wood are 50 DM/t and proceeds for contaminated wood (coated timber) 150 DM/t (both 50 per cent of the input material). See http://www.google.co.in/url?sa=t&rct=j&q=m.%20specht*%2C%20a.%20bandi*%2C%20f.%20baumgart*%2C%20c.%20n.%20murray**%2C%20and%20j.%20gretz***%20%20synthesis%20of%20methanol%20from%20biomass%2Fco2%20resources%20&source=web&cd=1&ved=0CEYQFjAA&url=http%3A%2F%2Flampopumput.info%2Ffoorumi%2Findex.php%3Faction%3Ddlattach%3Btopic%3D4175.0%3Battach%3D6781&ei=2Mk0UOLFNYX-IrQfJ5IHgDA&usg=AFQjCNH1cB80h-SsHP40JVsJZImPSlEe3Q (accessed on 11 July 2017).

[13] Richard L. Bain, 'Material and Energy Balances for Methanol from Biomass Using Biomass Gasifiers' (Golden, CO: National Renewable Energy Laboratory, 14 January 1992).

14   P. Galindo Cifre and O. Badr, 'Renewable Hydrogen Utilisation for the Production of Methanol', *Energy Conservation and Management* 48 (2007): 519–527.

15   Tomas Ekbom, Mats Lindblom, Niklas Berglin and Peter Ahlvik, 'Technical and Commercial Feasibility Study of Black Liquor Gasification with Methanol/DME Production as Motor Fuels for Automotive Used—BLGMF' (EU ALTENER project, 2003 December), Report for Contract No. 4.1030/Z/01-087/2001, 95–96.

16   Ibid. See also T. Ekbom, N. Berglin and S. Logdeberg, 'High Efficient Motor Fuel Production Biomass via From Black Liquor Gasification'.

17   *Earth Times*, 'NewPage to Discontinue Building Biofuels Plant with Chemrec AB of Sweden' (17 July 2009).

18   *Greenaironline.com*, 'Major Study into Feasibility of a Jet biofuel Refinery at Stockholm–Arlanda Provides Encouragement', available at http://www.greenaironline.com/news.php?viewStory=1112 (accessed on 22 March 2014).

19   Volvo Group, 'Promising Future for Bio-DME as Vehicle Fuel', available at http://www.volvogroup.com/group/global/en-gb/newsmedia/corpnews/_layouts/cwp.internet.volvocom/newsitem.aspx?news.itemid=120233&news.language=en-gb (accessed on 30 August 2012).

20   See http://www.topsoe.com/sitecore/shell/Applications/~/media/PDF%20files/Topsoe_Catalysis_Forum/2008/Klintbom.ashx (accessed on 3 September 2012).

21   Pam Radtke Russel, 'Changing Economics Improve Plasma Technology's Outlook', *Engineering News-Record* (ENR.com, 12 May 2007), available at http://www.enr.com/ (accessed on 1 September 2012).

22   With the onset of severe recession in 2008, many of these projects have also had to be postponed or abandoned.

23   See http://advancedbiofuelsusa.info/solena-sas-partner-for-aviation-biofuels-project-at-stockholm-airport (accessed on 11 July 2017).

24   See http://www.solenafuels.com/sites/default/files/Seven%20ATA%20Member%20Airlines%20sign%20LOI%20with%20Solena%20Fuels.pdf (accessed on 11 July 2017).

25   4,600 kcal/kg which is almost the same as that of a good hardwood.

26   B.M. Jenkins, L.L. Baxter, T.R. Miles Jr. and T.R. Miles, 'Combustion Properties of Biomass', *Fuel Processing Technology* 54 (1998): 17–46.

27 In 2011–2012, (see http://www.sugarnutrition.org.uk/the-produc-tion-of-sugar.aspx [accessed on 27 March 2014]), the availability of bagasse and sugar cane waste is estimated on the basis of a production of cane sugar of 137 million tonnes, a 12.5 per cent yield of sugar from the cane, one tonne of air-dried bagasse from four tonnes of sugar cane and one tonne of sugar cane waste from one tonne of sugar cane.

28 Details of the basis for this estimate are given in Chapter 7.

29 See http://doctor-biochar.blogspot.in/2013/10/structural-and-chem-ical-composition-of.html (accessed on 26 January 2017).

30 K. Jindo, H. Mizumoto, Y. Sawada, M.A. Sanchez-Monedero and T. Sonoki, 'Physical and Chemical Characterization of Biochars Derived from Different Agricultural Residues', *Biogeosciences* (2014 December), available at http://www.biogeosciences.net/11/6613/2014/bg-11-6613-2014.pdf (accessed on 11 July 2017).

31 Ibid.

Chapter Eight

Completing the Energy Shift—Hydrogen from Sunlight

This chapter examines the current rate of addition of carbon to the atmosphere and the increases in the pipeline, and concludes that while the technologies described in the previous four chapters can bring annual emissions down close to zero before the end of the century, they will not be able to remove the substantial carbon overhang that will have developed by the middle of the century. It describes a revolutionary technology for extracting hydrogen from water using only sunlight, and combining it with CO_2 extracted from the air that is nearing commercial viability in the United States.

The last four chapters have explored in some detail the capacity of two major technologies—biomass gasification and CSP—to shift the energy base of mankind from fossil fuels to the sun. Three questions remain to be answered: how complete can the energy shift they make possible be? How quickly can the shift be made? And how much, over and above what the world is already spending on fossil fuels, is it likely to cost?

Power generation accounted for 47 per cent, that is, 17 billion tonnes of GHGs released in 2011, transport for 5.5 billion tonnes. And the petrochemicals and fertiliser industries for another 5 billion tonnes of GHGs. In all, therefore, the two technologies can reduce CO_2 emissions by 27 to 28 billion tonnes.

This is pretty much all of the anthropogenic carbon emissions that has been added to the atmosphere every year during the past decade.[1] Even if this could be done in one stroke—say by 2020—the concentration of CO_2 in the atmosphere by 2100 AD would still be around 410 ppm. So by the time the earth absorbs the full impact of this increase, say a hundred years later, its mean atmospheric temperature will have risen by at least 3°C or more, as it had done in the Permian age.

But a sudden, complete, cut-off of GHG emissions is impossible. On the contrary, the investments that global corporates and national governments are pouring into the discovery and extraction of fossil fuels will not only ensure a further rise in CO_2 concentration but ensure that it takes place faster and faster. Global warming will therefore accelerate. So even if the shift out of fossil fuels is completed by the end of this century, it will not suffice to keep the rise in air temperature below the 2°C target the world has set itself, let alone the 1.5°C that scientists consider safe. That is why both Christiana Figueres of the UNFCCC and the IPCC's fifth assessment report have concluded that governments must find ways to start taking carbon 'out' of the air as well.

Stabilisation of CO_2 emissions is not enough

In the industrialised countries, the campaign against fossil fuels has focused almost exclusively upon coal. Coal accounts for three-fifths of the total electricity generated in the world, and coal-fired power plants produce at least twice as much CO_2 per unit of electricity as natural gas. So every proposal for averting abrupt climate change has begun by asking governments to phase out its use. The United States is doing so by replacing coal-burning power plants with new ones that use its shale gas. But natural gas is still a fossil fuel and its supply is limited. Hence the frantic search for ways to sequester flue gas CO_2 underground.

Storage underground has proved a non-starter, but there is another, easier, way to achieve the same result. This is to convert the CO_2 into methanol. The proposal was put forward in 2005 by the 1994 Nobel Laureate in chemistry, George Olah, but while the chemistry of the conversion was well understood, it could not be made economically viable because of the prohibitive cost of producing hydrogen. That barrier has now been broken.

Hydrogen from the sun

In 2006, when President George Bush announced in his State of the Union address to Congress that he had opted for ethanol as the renewable transport fuel of the future, George A. Olah had just published a path-breaking book *Beyond Oil and Gas: The Methanol Economy*. In it, he and two colleagues Alain Goeppert and G.K. Surya Prakash had chalked out a grand vision of how humanity could duplicate photosynthesis in a highly accelerated form and suck carbon dioxide out of the air to bring its concentration back to safe levels.

Living biomass, they reminded their readers, breathes in carbon dioxide from the air during the day, uses the sun's energy to release hydrogen from its its own water and combines it with the CO_2 to produce more biomass. In the process, it exhales the oxygen released by the break-up of the water molecule. This replenishes the oxygen breathed in by animals and birds. Olah proposed duplicating, and vastly accelerating, this cycle by using electricity in place of sunlight to break down the water and combine the released hydrogen with CO_2 to produce methanol.[2]

The production of methanol from CO_2 is not a new technology. The first commercial plant to do this was set up in the United States in 1922. But the rapidly falling price and easy availability of natural gas in the 1920s and 1930s pushed this technology out of the market. In spite of that, in 2005, when Olah released his book, one power plant, located at Coleson Cove in New Brunswick, Canada, was doing precisely that in order to satisfy Canadian environmental standards.

Interest in capturing carbon by this route began to revive only after the establishment of the Clean Development Mechanism under the Kyoto Protocol in 1997. New flue gas cleaning technology had brought down the cost of obtaining pure CO_2 very substantially and was in use in many industries. Thus, when the EU put its CDM in place to reward enterprises that cut their CO_2 emissions, it provided a powerful stimulus to research centres and engineering faculties to find an economically profitable ways to convert it into transport fuels. When Olah's book came out, scientists in several universities had been studying the possibility of capturing carbon dioxide by converting it into methanol for some time.

In November 2000, D. Mignard, M. Sahebzada, J.M. Duthie and H.W. Whittington, four scientists from the faculty of engineering at the University of Edinburgh, the Heriot-Watt University in the same city and the Imperial College London, carried out a detailed economic appraisal of the cost of producing methanol from flue gas CO_2 under different assumptions of the way the plant would be run. Based upon the typical load variations in a 1,000 MW power plant, they assumed that it would run at full load for 8 hours, would have 100 MW to spare for 12 hours and 500 MW to spare for the 4 lean hours of the night and calculated the cost of producing methanol (a) when the plant used all of its spare generating capacity, as and when it was available to produce hydrogen and (b) when it used only the 100 MW that was available for 16 hours and bought 100 MW of power from the grid for the remaining 8 hours.

They found that option (a) was uneconomic because it required the installation of a 500 MW electrolysis unit that would run at its full capacity for only 4 hours a day. This cost four times as much as a 100 MW electrolyser and pushed up the capital cost of the plant by three and a half times, but yielded only twice as much methanol as the second option. Mignard and his colleagues estimated that in November 2000 the second option would allow mithanol to be produced at 27 pence per gasoline litre equivalent (GLE). In November 2000, its price at the pump was 84.4 pence.

Thus, the technology would be economically viable if the government simply exempted it from some of the taxes it was levying on gasoline.[3] But the UK government showed no interest in doing so. Nor, in contrast to the billions that it poured into the storage of CO_2 underground, did it offer to subsidise a trial in a single British power plant.

Instead, the implicit demand of all governments throughout the dozen years that have followed the publication of the Edinburgh paper has been the same: any technology that hopes to replace gasoline *must be cheaper than its 'bare' cost to the oil marketing companies.* None of the European governments, or the state governments in the United States that levy hefty sales taxes on gasoline, has even considered reducing these to give this technology a chance. As is shown in Chapter 9, this is not mere oversight.

The hurdle to bringing down the cost of producing methanol from flue gas down to the bare cost of gasoline has been the vast amount of electricity that is needed to produce the required hydrogen. To break the water molecule and produce 1 kg of hydrogen requires 32.935 KWh of power, but energy leakages, mostly in the form of heat, raise this further. In spite of this, the possibility of capturing at least some of the flue gas emitted by coal-fired power plants has continued to fascinate researchers all over the world. In 2008, two US-based companies QuantumSphere and DoppStein announced that using a three-metal, nanoparticle-coated electrode that increased the surface area in contact with the water by as much as 1,000 times, they had speeded up the rate of generation of hydrogen at ambient temperatures and pressures by more than 20 times. This promised to reduce the size, and therefore capital cost, of electrolysis plants dramatically.[4]

The focus of still more recent research has been on finding a way to circumvent, or minimise, the need for electrolysis. At the Institute of Bioengineering and Nanotechnology, Singapore, scientists have developed a process for combining CO_2 and water without using expensive catalysts.

In Chennai, India, two scientists Christopher Kanakam and K. Praneeth Raj—one a chemist and the other an engineer—suggested another low-cost method of bypassing the need for electrolysis. This was to scrub the flue gas of sulphur dioxide, particulates, tar and other trace impurities by passing it through water[5] and then pass the moist gas over a bed of heated coke at around 1,000°C. The residual moisture, and the small amount of oxygen contained in the flue gas, is sufficient to force open the bond of carbon and oxygen in CO_2, convert the bulk of it into CO and produce a little hydrogen. The CO can then be converted via the 'water gas shift reaction' to give the desired proportion of hydrogen for the production of olefins and transport fuels. Alternatively the CO can be fed back into the power station's gas turbines to produce more power.

Kanakam and Raj estimated that doing this would reduce the coal required for power generation by half. What is more, because the chemical reactions involved generating more heat than they absorb, once the coke bed is heated to the required 1,000°C the process will need very little added energy to stay at that temperature.[6]

Hydrogen from sunlight

The most momentous recent development has been the development of a way to produce hydrogen from sunlight. In February 2012, a California-based company HyperSolar Inc. signed an agreement with the University of California, Santa Barbara (UCSB), and gave an end-to-end laboratory demonstration of two prototypes in July.[7] A key challenge it had met was to produce nanoparticle-sized, photo-voltaic film-coated, electrolysis units that would contain both the positive and negative terminals through which electricity generated by the PV film would pass to break the bond of the water molecule. The challenge they overcame was to make these out of materials that would not corrode rapidly when immersed in an acidic electrolyte.

Since the original demonstration, progress has been rapid. In October 2012, the company and UCSB filed a patent application for the production of a low cost nanocell that did not get corroded and did not require the use of expensive platinum catalysts to produce hydrogen. In April 2013, the process received the endorsement of the American journal *Nano Letters*. The reviewers concluded that there were no obstacles to scaling up the production of hydrogen by this route.[8] By March 2014, HyperSolar had developed its cells to the point where they were yielding a voltage of 1.2 volts, just 0.3 volts short of the minimum potential difference between the positive and negative terminals needed to break water molecules into hydrogen and oxygen.

HyperSolar crossed this boundary in September 2015 when it achieved a potential difference of 1.55 volts.[9] On 1 December 2015, it succeeded in replacing platinum with an inexpensive, sulphide-based catalyst with a life of 'several hundreds of hours'. In July 2016, the company announced that it had nearly tripled the volume of hydrogen being produced by the current it was generating and had achieved two-thirds of the 10 per cent solar energy-to-hydrogen-conversion efficiency that it considered necessary to make the process commercially viable.[10] In November, it released a video which showed that the process was more efficient with contaminated than pure water. The commercial barrier to producing hydrogen from sunlight is therefore on the point of being breached.

The impact that this will have on the energy industry is mind-boggling. Carbon capture by this route will not prolong coal-based power generation indefinitely. But it will offer the 450 or more coal-fired plants that have been set up around the world in recent years to choose the pace at which they will exit from the industry. For coal miners and mining townships, it offers a gentler transition into the future. To China, India, Russia, Poland and Brazil, countries with large reserves of coal, it offers a way to prolong its use while bringing down their CO_2 emissions. In sum,

HyperSolar's technology not only offers a way to avert the head-on conflict between environment and market that has stalled the shift out of fossil fuels, but breaks the last remaining hurdle to limiting the rise in global temperatures to 2°C, and possibly even 1.5°, by extracting large volumes of CO_2 from the air and turning it into transport fuels and olefins.

Beware of the market

If it is deployed rapidly enough, say by the middle of this century or a little later, the world will not only have stopped adding to the CO_2 in the atmosphere but will be well on its way to reversing the 'overshoot' in concentration that is bound to take place in the next four decades. But this will happen only if the governments of the major economies jointly decide not to leave its dissemination at the mercy of the market and decide to support all the companies that adopt it with tax breaks and subsidies. Unfortunately the market is once again all set to drag the world in the opposite direction.

In the last week of January 2017, General Motors (GM) and Honda announced a partnership in which each company will contribute $85 million to develop a fuel cell system manufacturing facility located at GM's electric vehicle battery site in Michigan. Expected to break ground in 2020, the new factory is designed to meet corporate and international expectations that hydrogen fuel cell vehicles will become a commercially viable alternative to gasoline-powered vehicles. The announcement comes following the formation of the Hydrogen Council, a group of 13 companies, that includes auto manufacturers Honda, BMW, Toyota, Hyundai and several industrial oil, gas and mineral companies, whose stated purpose is to 'convince other companies, regulators and the public that switching to hydrogen is essential to our continued survival'.[11]

The formation of the Council was announced at the World Economic Forum at Davos in 2016. Its purpose was to spur consumers' acceptance of hydrogen-powered automobiles and therefore a rapid spread of hydrogen-fuelling stations. A report prepared by DC-based *Information Trends* estimates that there will be 5,000 hydrogen-fuelling stations worldwide by 2032, a tremendous uptick from the existing 285 in 2016.[12]

The Council may seriously believe that it is trying to save the world, but the effect of its advocacy of hydrogen 'will be the exact opposite'. By 2032, even in the best case scenario charted after the Paris summit, the atmospheric concentration of CO_2 will be closing in on 450 ppm, and three quarters of the UN-estimated maximum permissible addition of CO_2 to the air will have been exhausted, and 'we will still have close to a million petrol pumps to replace'.[13] Persuading the public to replace petrol- and diesel-powered automobiles with hydrogen fuel-celled vehicles will take even longer. By that time, climate change will be completely out of human control.

If this is to be prevented, the world needs to understand precisely what is feeding its death wish. Previous chapters have shown that the main driver has been a near-religious belief in the infallibility of the market economy and the need for the state to virtually abdicate and cede this responsibility to the market. Therefore, it is time to examine the foundations of this belief.

Notes and references

[1] See https://scripps.ucsd.edu/programs/keelingcurve/2013/05/20/now-what/ (accessed on 11 July 2017) and 'Summary for Policymakers', in IPCC's Fifth Assessment Report, 6–8. These and all other numbers for GHG emissions cited in this book are median estimates, without the uncertainty margins given by the IPCC out.

[2] George A. Olah, Alan Goeppert and G.K. Suryaprakash, *Beyond Oil and Gas*, 243–245.

[3] D. Mignard, M. Sahibzada, J.M. Duthie, H.W. Whittington, 'Methanol Synthesis from Flue-gas CO_2 and Renewable Electricity: A Feasibility Study'. *International Journal of Hydrogen Energy*, 28 (2003), 455–464.

4   'Highly Efficient Hydrogen Generation via Water Electrolysis Using Nanometal Electrodes'. Available at: http://www.jonblackmon.com/ Efficient_Electrolysis.pdf (accessed on 20 August 2017).
5   See table below:

Fossil Fuel Emission Levels—Pounds per Billion Btu of Energy Input

Pollutant	Natural Gas	Oil	Coal
Carbon Dioxide	117,000	164,000	208,000
Carbon Monoxide	40	33	208
Nitrogen Oxides	92	448	457
Sulfur Dioxide	1	1,122	2,591

Source: Kanakam and Raj (2011). Also see note 6.

6   Christopher Kanakam and K. Praneeth Raj, 'Recycling Carbon Dioxide to Provide Renewable Energy and for the Mitigation of Greenhouse Gas Emissions', *2011 2nd International Conference on Environmental Science and Technology IPCBEE* 6 (Singapore: IACSIT Press, 2011), available at http://www.ipcbee.com/vol6/ no1/58-F00115.pdf (accessed on 11 July 2017). The water gas shift reaction is $CO + H_2O = H_2 + CO_2$.
7   Two videos of the demonstration are available on the Internet at http://www.hypersolar.com/ (accessed on 23 September 2012).
8   Syed Mubeen, Nirala Singh, Joun Lee, Galen D. Stucky, Martin Moskovits and Eric W. McFarland, 'Synthesis of Chemicals Using Solar Energy with Stable Photo Electrochemically Active Heterostructures' (Santa Barbara: Department of Chemistry and Department of Chemical Engineering), *Nano Letters* 13, no. 5 (2013), 2110–2115. The abstract of their finding is reproduced below:

Efficient and cost-effective conversion of solar energy to useful chemicals and fuels could lead to a significant reduction in fossil hydrocarbon use. Artificial systems that use solar energy to produce chemicals have been reported for more than a century. However the most efficient devices demonstrated, based on traditionally fabricated compound semiconductors, have extremely short working lifetimes due to photocorrosion by the electrolyte. Here we report a stable, scalable design and molecular level fabrication strategy to create photoelectrochemically active heterostructure (PAH) units consisting of an efficient semi-conductor light absorber in contact with oxidation and reduction electrocatalysts and otherwise protected by alumina. The functional heterostructures are fabricated by layer-by-layer, template-directed, electrochemical synthesis in porous anodic aluminum oxide membranes to produce high density arrays of electronically autonomous, nanostructured, corrosion resistant, photoactive units ($\sim 10^9$–10^{10} PAHs per cm^2). Each PAH unit is isolated from its neighbor by the transparent electrically insulating oxide cellular enclosure that makes the overall assembly fault tolerant. When illuminated with visible light, the free floating devices have been demonstrated to produce hydrogen at a stable rate for over 24 h in corrosive hydroiodic acid electrolyte with light as the only input. The quantum efficiency (averaged over the solar spectrum) for absorbed photons-to-hydrogen conversion was 7.4% and solar-to-hydrogen energy efficiency of incident light was 0.9%. The fabrication approach is scalable for commercial manufacturing and readily adaptable to a variety of earth abundant semiconductors which might otherwise be unstable as photoelectrocatalysts.

[9] Here is the text of the e-mail HyperSolar sent to its correspondents:

SANTA BARBARA, CA—September 15, 2015—HyperSolar, Inc. (OTCQB: HYSR), the developer of a breakthrough technology to produce renewable hydrogen using sunlight and any source of water, today announced that it had reached 1.55 volts (V), a significant milestone achievement in its effort to split water molecules for the production of renewable hydrogen fuel.

While the Company had previously met the challenge of exceeding 1.23 V, the theoretical minimum voltage needed to split water molecules into hydrogen and oxygen, a minimum of 1.5 V is needed to do so for commercially viable real world applications. Examples of these commercial applications include hydrogen-charging stations for fuel-cell vehicles, or warehouse transportation for big box retailers. The 1.55 V breakthrough, in a low-cost single solar cell element, is representative of the tremendous progression of the technology, as the Company announced its 1.4 V breakthrough just one month ago. The results were recorded at the University of Iowa campus research location where researchers are currently focused on further increasing the voltages and currents achievable from their inexpensive light driven hydrogen generation particles.

Following this breakthrough, the company will focus its efforts on increasing the hydrogen production efficiencies of these particles by bonding the ideal fuel production catalyst to the low-cost high-voltage solar cell. In order to achieve this, the company is currently exploring two parallel approaches. The first is to identify materials that interface with well-known hydrogen production catalysts, such as the platinum on solar particles, to improve sunlight-to-hydrogen conversion efficiency. The second is to pursue methods that further increase photo voltages of solar particles to greater than 1.7 V that allow integration of cheaper earth abundant catalysts without significant loss in hydrogen production efficiency.

'This announcement represents one of the most important milestone achievements the Company has made to date', said Tim Young, CEO of HyperSolar. 'Both the University of Iowa and University of California, Santa Barbara teams have been instrumental in spurring the speed at which our technology has developed, resulting in this voltage breakthrough. We are focused on identifying the next steps for the technology that will make it possible for us to scale up to make a commercial technology that can produce hydrogen fuel at or near the point of distribution, using only water and sunlight'.

10 See https://mail.google.com/mail/u/0/#search/hypersolar/1515ce-9035cba5af (accessed on 11 July 2017).

11 See https://www.engadget.com/2017/01/17/auto-and-oil-industries-want-to-pit-hydrogen-cars-against-evs/ (accessed on 17 February 2017).

12 See http://www.informationtrends.net/hydrogen-fueling-stations.html (accessed on 8 February 2017).

13 In 2014, eight countries alone accounted for nearly half a million petrol pumps. These included the United States, Canada, Argentina, Russia, China and India, but only Germany and the United Kingdom from the EU. Another half million for the remaining 180-plus countries seems a reasonable estimate. See https://en.wikipedia.org/wiki/Filling_station (accessed on 11 July 2017).

Part III

The Market Economy Is Not a Friend

Chapter Nine
The Inertia of the Market Economy

If so many renewable energy technologies are already viable, or close to becoming so, why does the world know so little about them? The answer lies in the way an oligopolistic global economy actually functions, against the way it is supposed to function.

When so many technologies have been developed that can produce power, transport fuels and petrochemicals, and some are already doing so, why do we know so little about them? Why did Renault's recent launch of its newest all-electric car, the Zoe, receive nationwide publicity in France despite the fact that electric cars have not received the welcome from consumers that their manufacturers had hoped for?[1] And why does one not come across any mention in the voluminous writing on 'green' technologies of the fact that in California automobiles ran on methanol-blended fuels for 2 million miles in the 1990s and early 2000s, before being abandoned in favour of ethanol?

This has been the fate of the entire array of the emerging and established technologies described in the preceding chapters. Had these been developed with the same zeal that electric cars, hydrogen fuel-cell engines and fusion power are being developed today, the threat from global warming might have been contained, if not overcome, by now. If they are adopted in the next two decades, they can still limit the damage that the present concentration of CO_2 and other GHGs will do in the future and slow down its development sufficiently to enable humanity to adjust to it in peace. Why, then, have they been so studiously ignored?

The answer lies in the resistance that the modern market economy puts up to fundamental shifts in technology. Despite being primarily climate scientists, the 2,500 who met at the Copenhagen Science Congress in 2009 understood this. In their memorandum to the Danish prime minister, they wrote:

> To achieve the societal transformation required to meet the climate change challenge, we must overcome a number of significant constraints and seize critical opportunities. These include reducing *inertia in social and economic systems*; building on a growing public desire for governments to act on climate change; removing implicit and explicit subsidies; reducing the influence of vested interests that increase emissions and reduce resilience.... (Emphasis added)[2]

This resistance is rooted in the desire of investors to slow down the pace of technological change in order to minimise the obsolescence of their current investments. But the idea that markets actually resist innovation goes against the idealised picture of their functioning that neoclassical economics has painted. In this picture, markets foster competition; competition rewards the efficient and penalises the inefficient and eventually drives the latter out of business. This process of weeding out leads to constant improvements in quality and productivity, and consequently, to increases in prosperity.

What economic theory relegates to the background is the unpalatable fact that the weeding out also leaves fewer and fewer producers in the market. Competition therefore contains within itself the seeds of its own destruction. A truly competitive market is a transient stage in the development of the Market Economy.

The end product of unrestrained competition is oligopoly, if not outright monopoly. The fewer the producers that remain in the market, the larger is their scale of production and, therefore, the loss investors will sustain if a new technology makes their existing production processes obsolete. The longer they can stay in production, however, the more can they recoup their investment. Thus, as competition shades into oligopoly, the stake of established producers in slowing down the pace of technological change grows steadily larger. Oligopolistic markets are, therefore, characterised by a constant struggle between those who seek to introduce new technologies and those who seek to prevent or at least delay it. Schumpeter called this struggle between giants 'creative destruction'.[3]

Over the past century, the attempts to slow down the pace of change of technology have taken several clearly recognisable forms. First, producers seek to ward off the threat from new technology by constantly improving the quality and versatility of their existing products and processes. This is an incremental process that promises great benefits at relatively little cost because it does not render their existing investment obsolete. So it is heavily, sometimes frantically, promoted. The relentless promotion of energy-efficient models and technologies is a good example.

A second form of resistance is to buy out the inventors of new products and processes that, in the hands of their rivals, could threaten their position in the market. This has given birth to a thriving market in patents and copyrights. A third is to develop or modify new products or processes of their own to pre-empt the success of their rivals' products. Competition in oligopolistic markets therefore takes myriad forms from concentrating resources on specific fields of research to buying the patents of competing products, to indulging in industrial espionage.

There is, however, a fourth route that is open only to very large, powerful enterprises, such as the global corporations of today, and to well-organised industrial lobbies. It is to enlist political power to protect their investments and stall technological change. This too has taken several forms, such as getting governments to subsidise existing products and processes while denying fiscal and other start-up incentives to new ones; refusing to create the infrastructure needed by new technologies at public cost and denigrating, or sabotaging, programmes designed to prove the reliability of a new technology.

Going 'green' is 'greenwash'

All of these stratagems have been used by corporations, industry associations and lobbyists to reinforce the belief that there is no alternative to fossil fuels. The promotion of energy saving as the most effective way to combat global warming is reflected by the flood of 'green' goods, energy-saving devices and bio-organic foods and lotions released by the signing of the Kyoto Protocol. Relentless advertising has given the words 'green' and 'bio' a feel-good association that makes people believe they are doing their bit to avert global warming when they buy goods that sport these labels. Plastering the walls and roofs of residences and office buildings with PV cells and claiming tax benefits for doing so is another example of a feel-good technology that, in the end, does very little to slow down the concentration of CO_2 in the atmosphere.

In the realm of manufacture, more firms are vying with each other to introduce innovations in the way they run their businesses that make them 'carbon-neutral'. A short list of strategies for a 'zero-carbon' society, cited with approval by the Stern Review Committee on Climate Change in 2005, reads as follows:

1. Toyota aims to build recyclable cars with zero emissions by minimising the environmental impact of vehicles over the life cycle of a car. Energy use can be reduced through efficient manufacturing and production, engine types offering potential to reduce emissions from driving and disposal at the end of life has been part of their vision of sustainable mobility.

2. In 2002, Avis Europe introduced a scheme to allow their car hire customers to offset carbon emissions, in partnership with the CarbonNeutral Company (formerly Future Forests). They state that they have become 'carbon-neutral' by 2005 by using their buildings more efficiently, recycling materials and off-setting non-reducible emissions via tree planting and support of renewable energy and technology projects to reduce GHG emissions.

3. Vattenfall, an energy company that operates hydro, nuclear and coal generators, has been developing and implementing three main CO_2-reducing measures: optimisation of existing technology to reduce emissions per unit of energy, increased use of non-CO_2 energy sources and a long-term project to capture and permanently store CO_2 from fossil fuel power plants.

4. Alcan has an ambition to become 'climate-neutral' by no later than 2020 through the full life cycle of its aluminium products. They have sought to increase energy efficiency through continued research and development in technology and process improvements, as well as reducing GHG emissions related to energy use, and pursuing the best energy mix from available energy resources and non-carbon-based energy projects.

5. HSBC became the world's first major bank to become 'carbon-neutral' in December 2005. To meet this goal, a carbon management plan has been put in place which consists of three parts: reducing direct emissions, reducing the carbon intensity of the electricity used by buying from renewable sources where feasible and offsetting the remaining CO_2 from the bank's own operations by buying emission reductions from 'green' projects.[4]

Today, Lord Stern is among the first to concede that such measures are simply not enough. On the contrary, by holding out the promise of arresting global warming they may have blunted the edge of the public's anxiety and weakened its will to act.

Chasing over-the-horizon technologies

Another method of stalling the development of readily deployable renewable energy technologies has been to vigorously promote futuristic, 'over-the-horizon' technologies that will take so long to develop that they pose no threat to existing investments in fossil fuels. Hardly a year passes without reports about dramatic break-throughs in energy technology that will resolve the problem of global warming virtually overnight. These share several common characteristics: they involve dramatic departures from existing technologies; they are at the laboratory or at most pilot demonstration stage of development and will therefore take decades to commercialise and all are ruinously expensive. The promoters, who are very often the companies that are most heavily invested in the current generation of fossil fuel technologies, assure us, however, that large-scale production will one day make these technologies economically viable. In the meantime, in an ironic reversal of the infant industry argument, they demand heavy subsidies during the scaling-up period.

Three such technologies, CCS, hydrogen-fuelled automobiles and cellulosic ethanol, were described in Chapter 2. So far, none, except a lone cellulosic ethanol plant in Brazil, has come even close to commercial viability. But a decade has been lost chasing these will-o'-the-wisps in the process. Today, another technology of questionable value is again raising the hope of a quick fix for global warming. This is the electrically powered motorcar or motorcycle. The technology is almost as old as electricity itself. What has always made it attractive to consumers is the possibility of recharging one's vehicle every night at home for next to nothing, while off-peak tariffs for electricity apply. In recent years, the mounting panic over global warming has created another justification for exploiting this technology: Since coal-fired boilers can only be banked, not shut down completely, during the dead of night when the demand for electricity is at its lowest, the increase in emissions from increasing electricity generation to recharge batteries will be considerably lower than the emissions from the transport fuels they replace. As a result, there is now hardly a car maker left in the world who is not marketing, or has plans to market, electric cars and two-wheelers.

The flaw in this justification is that it holds true only so long as the world continues to burn coal to obtain base-load power, and the number of vehicles needing daily recharge remains relatively small. Once power has to be generated from scratch to charge the batteries, the second law of thermodynamics will kick in: only 40 per cent of the energy contained by coal will get converted into electricity and a fraction of the electrical energy will be lost in its conversion to mechanical energy. In the end, without a breakthrough in providing renewable energy electric cars will not only prolong the use of coal but increase emissions instead of reducing them.

Time has become the enemy

The acceleration of global warming since the 1980s, and particularly since 2011, is a warning that the world may not have even the 20 years left to prevent it from taking on a life of its own. At the very least it is a warning that we can no longer afford the luxury of pinning our faith on unproven technologies. Despite this, it is only in such 'over-the-horizon', unproven, technologies that big business is still predisposed to invest. When the ethanol mania hit the United States in 2005–2006, a number of scientists expressed grave doubts about the feasibility of making the production of cellulosic, as opposed to corn-based, ethanol economically viable.[5] But these did not prevent industry from rushing in to finance research on cellulosic ethanol. Richard Branson of Virgin Atlantic pledged $3 billion to research on cellulosic ethanol. The British oil giant BP pledged a billion dollars. And the wholehearted involvement of the biggest industrial giants of the world persuaded Bush to pledge $400 million to research on the project. 'It's the holy grail ... *if you can make it work*', said John Felmy, chief economist at the American Petroleum Institute (API)[6] (Emphasis added).

Industry's fascination with unproven and untested technologies serves the interests of big business in three distinct ways. The first is that at a time of prolonged industrial slackness, their attempts to develop evolutionary technologies burnishes their image, while the subsidies governments offer boosts their bottom line: Honda hoped to cash in handsomely on its fuel cell-powered car even though it knew that it would never be anything more than an expensive novelty. Alstom, and a dozen other transnational corporations, gathered billions in grants and subsidies to fund their experiments in CCS underground, in the teeth of a veritable mountain of evidence that CCS would nearly double the cost of electricity and meet tooth and nail resistance from the affected communities at every turn. To overcome this, they warned their prospective clients that the cost to society will be far greater if the CO_2 is not sequestered in this way, unmindful of the irony that this is the argument they oppose when it is used to justify subsidising 'expensive' renewable energy technologies. The role of the US farm lobby in pushing for a subsidy on corn-based ethanol has been described earlier, and more than a dozen companies have garnered close to a billion dollars with promises to produce cellulosic ethanol that they have not yet fulfilled.

Misdirection—The dark side of business as usual

Side by side with promoting red herrings, big business is using the stratagem of deflecting attention from proven technologies that can be adapted to facilitate a rapid shift out of fossil fuels with a minimum of research to futuristic technologies that will pose no threat to its existing investments. It is no coincidence that all the four red herrings described above are to be found in the realm of transport fuels: even the votaries of CCS are talking only of storing the CO_2

underground, and not the far cheaper technology for converting flue gas CO_2 into transport fuels that has been known since 1922 and is in use at Coleson Cove, New Brunswick, in Canada today.

That this misdirection may not be accidental became apparent in 2009, when the Obama administration tried to push the Waxman–Markey Bill to limit CO_2 emissions in the United States through the US Congress. To defeat the bill, the API sent an e-mail to its members asking them to organise a nationwide 'astroturfing' campaign designed to create the impression that there was widespread grassroots opposition to the bill. This was leaked by one of the API's members, who opposed the tactic, to Greenpeace. The API had asked its members to organise a series of 'energy citizens' rallies' on the lines of the republicans' 'Town Hall' meetings against the Health Bill. According to a report posted on the website 'BusinessGreen', in his e-mail, API President Jack Gerard urged the group's member companies to encourage staff to attend the planned rallies and to also extend invites to 'all vendors, suppliers, contractors, retirees and others who have an interest in our success'. He stressed that attendees would have to do little more than turn up. API would provide all the 'upfront resources' and had appointed 'a highly experienced events management company that has produced successful rallies for presidential campaigns, corporations and interest groups' to manage the events. The e-mail also called on member companies to not disclose details of the planned events, urging them to 'please treat this information as sensitive and ask those in your company to do so as well ... we don't want critics to know our game plan'.[7]

According to Greenpeace, by August 2009, the oil industry had raised $34 million to fight the Waxman–Markey Bill.[8] The very next year, undaunted by its exposure in the media, the oil industry again took the lead and collected $9 million of campaign funds to lobby for the revocation of a bill passed in 2006 by the California state legislature to cut emissions by 15 per cent by 2020. The law was intended to go into effect in 2012.[9]

The burial of methanol

Encouraging a perennial quest for new technologies is not the only method that big business has used to postpone the shift out of fossil fuels. The other method has been to systematically discourage the development of alternatives. One that oil and gas companies have consistently tried to crush is methanol. As an alternative to gasoline, methanol was used long before ethanol. Interest in its use as an alternate fuel to gasoline had never quite died out after the Germans used it to fight in the Second World War. But except in South Africa where SASOL began to produce synthetic fuels from coal in 1955, and still meets almost half the transport fuel requirements of the country from them, it never took off. Over the past four decades, there have been several attempts to introduce methanol as an alternative to gasoline as a transport fuel in the United States, but all have failed. The widely accepted explanations are the sustained fall in oil prices after 1980 and the 'absence of a lobby' (comparable to the US farm lobby) that would push for its adoption as a transport fuel. But the absence of such a lobby has given full reign to the lobbyists of the oil and gas industry to oppose the use of methanol. Their opposition has taken many forms. The commonest of these is a relentless highlighting of the real or supposed deficiencies of not only methanol but also ethanol as a transport fuel. As far back as 1949, S.J.W. Pleeth wrote in his book *Alcohol—A Fuel for Internal Combustion Engines:*

> The bias aroused by the use of alcohol as a motor fuel has produced results in different parts of the world that are incompatible with each other. In general we can detect two schools of thought with regard to the use of alcohol as a motor fuel. Countries with considerable oil deposits (such as the United States) or which control the oil deposits of other lands (such as Holland) tend to produce reports antithetical to the use of fuels alternative to petrol; countries with little or no indigenous oil tend to produce favourable report. The contrast between the two cases presented is most marked: one can scarcely avoid

the conclusion that the results arrived at are those best suited to the political or economic aims of the country concerned, or of the industry which sponsored the research.

Constant repetition, with little and at best sporadic rebuttal, has turned these allegations into facts.

The oil and gas companies have also not hesitated to intervene directly to stop research that could throw up alternatives to oil-based transport fuels. In December 1973, within weeks of the first Organization of the Petroleum Exporting Countries (OPEC)-engineered oil price rise, two MIT scientists Thomas B. Reed and R.M. Lerner published a paper in *Science* titled 'Methanol: A Versatile Fuel for Immediate Use'. In it, they pointed out that up to 15 per cent by volume of methanol could be blended with gasoline and used as a transport fuel without making any change in the settings of a typical motor car engine. Far from causing any loss of power, the addition of methanol reduced fuel consumption by 5 to 13 per cent and increased acceleration by up to 7 per cent. It did this in spite of the fact that methanol contained slightly less than half the energy of gasoline. Reed and Lerner went on to point out that methanol was already being produced in large quantities for industrial use from natural gas, but could be obtained in large quantities by gasifying biomass, such as urban solid waste, with oxygen to produce synthesis gas. This could be converted to methanol with catalysts that were already in use in industry.

The article was summarised in the *New York Times* and created a great deal of interest. This led to a private grant to MIT, where the two scientists were teaching at the time, of $100,000 to test the use of methanol as a blending fuel in motor spirits. But just when the scientists were about to run tests with 200 privately owned vehicles in Cambridge, Massachusetts, MIT cancelled the project and appropriated the unspent half of the project money for other uses. Reed left MIT in 1977 to work at the newly founded Solar Energy Research Institute at Golden, Colorado.

On 'Woodgas', the website of the Biomass Energy Foundation that Reed went on to establish in 1997, he later described the series of events of 1973 as follows:

> I became interested in the question of alternate fuels in 1973 when my boss asked me to look into 'hydrogen from sunlight' as a possible new energy source. However, hydrogen does not occur in nature and Mother Nature has had 3 billion years of development on renewable fuels through photosynthesis. So I began inquiring about alcohols as fuels ...
>
> ... Methanol is by far the simplest and cheapest ($0.15/gallon at that time) synthetic fuel to make synthetically. It is used in preference to gasoline at the racetrack, and I learned in 1973 that student teams had converted cars to methanol and won clean air car races ...
>
> A close friend of mine, Dr John Anderson, had just patented a process for converting municipal waste to synthesis gas which could be used to make methanol. I wrote a draft of the article for *Science* magazine proposing that methanol would be the most likely replacement for gasoline as the oil ran out. I gave the article to another friend, Dr R.M. Lerner, to read and comment. He liked the article, but suggested that the transition to a methanol fuelled future would be easier if it could be blended in increasing amounts into gasoline, say starting with 5 to 10% and gradually developing more capacity as oil became increasingly difficult to find.
>
> I liked this suggestion, so tested a little methanol in gasoline and found it dissolved at all ratios (at room temperature). Being adventurous and having an aging 1969 Toyota, I got a gallon of methanol from the stockroom and added it to my tank to get a 10% mixture. It started fine and I spent the afternoon doing several errands—didn't notice any difference in the driving. After a few more tankfuls of 10% methanol in my gasoline, I noticed that the 'tank to tank' fuel economy was equal to or better than gasoline. As a chemist, I knew that a gallon of gasoline had twice as much energy as a gallon of methanol, so I didn't believe it, (and) decided to make some more scientific tests. I obtained a fuel burette (graduated cylinder) and mounted it in

the window of my car and drove over a test course involving a 1 mile hill climb using 0%, 5%, 10%, 15% and 20% methanol in the burette. My mileage went up from 26.8 MPG on gasoline to 28.8 MPG on 15% methanol (a 7% increase), then declined but was still drivable.

Unfortunately the oil and motor industry was less than enthusiastic. A few months later a grant of $1,000,000 was given to the MIT Energy Laboratory. We were informed that the oil companies were the real experts in this field and so the permission to run the tests was revoked after we had spent 6 months and $50,000 developing our test methods.[10]

In 1975, *Science* sent a reporter Allen L. Hammond to MIT to investigate the project's cancellation. Hammond's article appeared in the November 1975 issue of the magazine.[11] In it he concluded that the $1,000,000 grant that MIT received had been given in equal parts by Exxon and the Ford Motor Corporation, and that the influence these companies acquired may have led the head and administrators of the Energy Research Laboratory at MIT to cancel the project.

Hammond unearthed no direct evidence of strings attached to the $1 million grant, but he found that MIT had just set up the Energy Research Laboratory and was in search of funds; that the funding had been obtained through active campaigning by the vice-president in charge of research at MIT; that 7 of the 24 members of its advisory board were from the auto and oil industries and that of the two scientists whose vote against the project proved crucial, one was the head of the automobile research section of MIT and the other was actually on loan from Exxon when he pronounced himself, as an expert, against the methanol testing project. Apparently he felt no qualms about a possible conflict of interest.[12] The abrupt cancellation of Reed and Lerner's project was not a lone intervention. A bill tabled in the California Legislature in 1975 that would have required methanol–gasoline blends to be sold in the state by 1980 was strongly opposed by oil and auto company spokesmen and eventually killed.[13]

Spain's retreat from solar energy

The oil lobby is not the only culprit. Given half a chance the coal, gas and nuclear power industries behave in exactly the same way. A prime example is the decision of the Conservative Government of Spain to withdraw subsidies from solar energy producers. One of its first victims is likely to be the Gemasolar plant, whose success in providing 24/7 power has electrified the climate change world. Since entering operation more than two years ago, Gemasolar has become the iconic plant of the solar power industry. In September 2013, Mercedes Benz used it as the backdrop in its advertising campaign to launch its new Sprinter van. Sony has used it in a new version of one of its most popular video games.

In 2011, Gemasolar won the first prize in the innovation category of the European Business Awards. And in September 2013, the International Federation of Consulting Engineers included it in the dozen projects it considered the best engineering works of the last century.

But far from taking pride in Spain's achievements in this field, the right-wing People's Party (Partido Popular in Spanish), which came to power in 2011, began doing away with all subsidies given to renewable energy producers to bridge the gap between their cost of production and the 'feed-in tariff' they could charge from the distribution companies. After calling a halt on all new renewable energy projects upon coming into office, in July 2013 José Manuel Soria López, Minister for Industry, Energy and Tourism, proposed an energy reform that would replace feed-in tariff subsidies with a 'reasonable rate of return of 7.5 per cent'. Spain is not alone: similar cuts are taking place all over Europe. The governments are justifying this by pointing out that the cost of producing solar energy has fallen sharply. This is true and explains why insurance companies are prepared to cover the risk of investment in solar power plants. But Spain has gone a step further and is imposing retrospective cuts in the subsidies that will force the solar power companies to shell out large sums to the government.

It will also make it virtually impossible for insurers to pick up the risk burden.[14]

The justification given by the Spanish government is that the debt the government has incurred through these subsidies has become too large to support. But among Spain's environmentalists it is common knowledge that the decision to stop subsidies altogether instead of tapering them off is being driven by Spain's five big utility companies, whose aim is to protect the investment they have made in combined cycle gas turbine generation during Spain's boom years. The connection surfaced when the energy bill presented to the Spanish Parliament by Soria Lopez was found to contain copy-and-paste references to a previous proposal from the energy utility Iberdrola.[15]

Most renewable power developers and operators expected to see cuts in subsidies after the change of government. But the degree of hostility shown to the sector has raised concerns that Spain's Ministry of Industry, Energy and Tourism is effectively in thrall to the country's big five utilities. Outwardly the big five are keen to push their green credentials, but within the energy industry they are known to be hostile to renewables because these are affecting the returns on their investments in combined cycle gas turbine plants during Spain's boom years. In November 2013, Termosolar Alcázar, a joint venture between Preneal of Spain and SolarReserve of the United States, announced that it was shelving plans for a 50 MW, €500 million CSP plant in Alcázar de San Juan, Castilla-La Mancha.[16]

Fracking and tar sands: Oil and gas go on the offensive

The prospects for shifting out of fossil fuels to power transport have also taken a turn for the worse. In the summer of 2009, the oil and gas lobby, which had been on the defensive so long as oil and gas prices had continued to rise, went on the offensive. The focus of mobilisation was Obama's introduction of the Waxman–Markey

Bill in the US Congress, but the underlying cause was a wholly unforeseen new development that was rapidly undermining the premises upon which big business' cooperation in the fight against global warming was founded. This was the onset of a third fossil fuel revolution. Until then no one had seriously doubted that the reserves of fossil fuels were running out. The known reserves of oil and gas were dwindling rapidly. Those of good quality coal were only a few decades behind. The world would, therefore, have to find alternatives to fossil fuels anyway. If doing so earlier rather than later also reduced the threat from global warming, then it would be killing two birds with one stone.

In 2008, this frail consensus was shattered by a technological breakthrough that promised a new age of plenty in fossil fuels. This was the hydraulic fracturing—nicknamed 'fracking'—of hard layers of shale deep underground to release vast amounts of shale gas and oil trapped in them. As the name implies, hydraulic fracturing involves pumping (shooting might be a more appropriate word) a cocktail of chemicals mixed with huge amounts of water laterally at very high pressures to soften and crack shale beds within the earth's crust. The same new technique of lateral drilling has also made it possible to skirt rock formations and reach tar sands and release billions of tonnes of oil trapped within them. These techniques have increased the US' output of natural gas from 49 billion cubic feet per day in 2005 to 74.1 billion cubic feet per day in 2015.[17] With an increase of 1 million barrels a day of 'tight' oil from the Alberta tar sands,[18] the new technology has already reduced the US' dependence on oil and gas imports from 60 per cent in 2007 to 46 per cent in 2011. Oil trapped in tar sands in Utah began to be extracted in 2016. The industry's spokespersons claim that there is an abundance of 'tight' oil and gas reserves in Pennsylvania, New York state, New England and Dakota in the United States and in Albert in Canada. A euphoric oil and gas industry is now promising future US governments a return to self-sufficiency in fossil fuels for North America for the next hundred years.[19]

Since the new technology surfaced in 2008, the quest for shale gas and 'tight' oil has spread to many parts of the world. The United Kingdom has found substantial reserves in Lancashire, China has found no fewer than seven shale beds with estimated reserves of 12 trillion cubic metres against the 9 trillion found in the United States and Israel claims to have found the largest single shale gas bed in the world off its shores in the Mediterranean. The US Energy Information Administration has estimated that 33 countries, including the United States, have 6,609 trillion cubic metres of recoverable shale gas reserves. This amounts to 40 per cent of the recoverable natural gas in conventional oil and gas deposits.[20] In the meantime, the melting of the Arctic has set off a race to find oil beneath the seabed. Oil companies have descended upon Greenland, where they suspect the presence of 20 billion tonnes of oil in offshore reserves which were hitherto locked under ice. Hectic prospecting for oil and gas has begun all around the Arctic shore. To quote Ed Luce of the *Financial Times*, 'A new era of fossil fuel appears to be upon us and nobody saw it coming'.[21]

The technology is spreading rapidly. Russian, Chinese, European and Saudi companies have joined the race to acquire it. New economic bonds are being forged between erstwhile adversaries who now realise that there is no God more powerful than money. Hence, Exxon has entered into an alliance with Russia's Rosneft to develop shale and tight oil reserves in Russia. Experts differ on whether fracking and the extraction of tight oil will bring down oil and gas prices, but all agree that it will slow down the rise and, in the case of gas, may even bring them down for a time.[22]

The opportunity for quick profits presented by 'fracking' and 'tight oil' is almost certainly the explanation for the sharp rise in lobbying activity by the oil and gas industry in 2008 and 2009. In 2008, oil and gas spent $178 million on lobbying the US Congress against the Waxman–Markey Bill. This was 10 times the sum that all the environmental groups and renewable energy companies together were able to spend. But this was not deemed sufficient, for in the first quarter of 2009, oil and gas' outlay rose by another 50 per cent.[23]

The determined opposition of this powerful sector of industry tilted the scales within the Obama administration. For while the arguments of the coal lobby were known, those of the oil and gas lobby—the promise of a return to the golden days of energy self-sufficiency, of millions of jobs, of lower energy prices and a restoration of America's political pre-eminence in global politics—were too seductive for any American administration to ignore. While the oil and gas industry are promising a return to energy self-sufficiency for the United States and hinting that this will restore its economic pre-eminence and political power, a series of studies have raised serious doubts about its safety. One single lateral well requires the injection of a minimum of 2 million gallons of water and 15,000 to 60,000 gallons of chemicals. Not only can this toxic mix get into underground reservoirs of drinking water, but when pumped out it has to be stored in huge ponds.[24] An attempt to pump this wastewater back into another hole in the ground is believed to be the most likely cause of an earthquake in an area of Ohio in which 'quakes had never been known before'.[25] There was a similar earthquake in Yorkshire, UK. The US Environmental Protection Agency has therefore refused to grant the shale gas industry special exemption from the Safe Drinking Water Act.

Discrediting the science

Despite these serious hazards, the allure of decades of easy, rising and assured profits from fracking and tight oil is proving irresistible. The energy lobbies have therefore redoubled their efforts to discredit the science behind global warming and impugn the reputation of the scientists whom the public trusts the most. To do the former, they have mobilised the dwindling group of scientists who still insist, against all the evidence to the contrary, that global warming does not pose a threat to humanity. The most eminent of them is Richard Lindzen, professor of meteorology at MIT.

Lindzen is among the very few climate sceptics who do not dispute the basic physics of global warming, first established through laboratory experiments by John Tyndall in 1880, that increasing the concentration of CO_2 in the air increases its capacity to trap heat and raises its temperature. But he claims that the models upon which scientists who predict doom if global warming is not arrested rely too heavily on one form of heat dissipation—through radiation. These models do not take into account the effect that cloud and water vapour formation, and high altitude air currents, have upon the pattern and rate of dissipation of the heat gathered from the sun. In an absolutely clear atmosphere—such as would have prevailed if earth had been a dead planet—the equilibrium temperature at which the earth would radiate all of the sun's energy back as heat would have been −18°C. But the actual temperature of the air on our planet is +15°. This is because there is already a greenhouse effect, and it is one that is necessary for the maintenance of life on earth. Clouds and water vapour are by far the most important regulators of this greenhouse effect. In fact, clouds reflect about 75 W/m² of the sun's energy back into space. Against this, even the doubling of CO_2 concentration in the atmosphere from 270 ppm in 1750 AD to about 550 ppm by 2050 AD will add only 2 W/m² to the earth's heat. Therefore, as Lindzen points out, it is perfectly possible for a small change in the absorptive and reflective capacity of clouds to neutralise the cumulative impact of 250 years of dependence on fossil fuels.

All this is uncontroversial. Climate scientists readily acknowledge the huge impact of clouds and water vapour and admit that this is the least understood part of the model they are trying to build. This gap in their models is why a majority remain cautious while drawing conclusions and bends over backwards to avoid making apocalyptic predictions.

But Lindzen is not content to stop here and feels no hesitation in going out onto thin ice. He has been insisting for more than 20 years that cloud and water vapour changes can only work in one direction—to dampen and not amplify human-induced warming. With dogged consistency, he continues to claim that 'recent studies of the physics of how deep clouds moisturize the atmosphere strongly suggest that this largest of the positive feedbacks (those that amplify warming) is not only negative (dampens it) but very large'.[26]

Twenty-two years later, the studies, he so blithely mentions, have still not seen the light of day. But in the meantime, Lindzen has refined his theory. He now claims that as the bulk of the increase in moisture caused by global warming takes place in the warm latitudes, a larger part of it will be precipitated as rain as the hot humid air rises, and a correspondingly smaller part will find its way into the upper atmosphere to be distributed as cirrus clouds that let the sun's energy in but trap more of as it heads out as infrared waves. The trouble with this theory is that, like the pro-catastrophe models he is criticising, his theory too only proposes a model. And that model too needs to stand the test of empirical scrutiny. Unfortunately the empirical studies intended to kick-start model building that have been carried out so far and suggest that the net effect of cloud behaviour is to amplify the initial stimulus provided by human activity.[27]

In the past decade, Lindzen has moved steadily further away from scientific scepticism towards metaphysical belief. On the day the Copenhagen Climate Conference opened, *The Wall Street Journal* (which also does not seem to see any danger in global warming) published an article by him titled 'The Climate Science Isn't Settled' in which he accused the scientific fraternity of dragging 'Our perceptions of nature ... back centuries so that the normal occasional occurrences of open water in summer over the North Pole, droughts, floods, hurricanes, sea-level variations, etc. are all taken as omens, portending doom due to our sinful ways'. In it, he also said repeatedly that there had been no warming of the planet during the previous 12 years despite a steady rise in CO_2 concentration, when all the data collected from thousands of observations all over the world confirmed the opposite.[28]

Despite this, Lindzen's unswerving belief in the stability of the climate equilibrium, his tenure at MIT and his solid reputation in the field of ozone chemistry have made him the perfect instrument in the hands of the business as usual lobby. His stature among the climate science sceptics was described recently by the *New York Times* as follows:

> At gatherings of climate change sceptics on both sides of the Atlantic, [he] has been treated as a star. During a debate in Australia over carbon taxes his work was cited repeatedly. When he appears at conferences of the Heartland Institute, the primary American organisation pushing climate change scepticism, he is greeted by thunderous applause.[29]

In 2003, Dr Lindzen was used by aides of President Bush to negate the findings of the Task Force on Climate Change of which James Hansen was a member, after a spate of letters from far Right Republicans in the Congress had made Bush renege on his promise to legislate a cap on CO_2 emissions.[30]

The euphoria generated by fracking has made global warming a taboo subject in affluent circles. Instead, well into 2015 the talk in the United States remained all about a return to 'energy self-sufficiency', the restoration of 'national sovereignty', a return to 'balanced' domestic and external budgets, the creation of hundreds of thousands, possibly millions, of high-paying new jobs and the return of 'American' money from distant lands to reap this new bonanza at home. How completely this had sidelined the effort to control climate change may be judged from the fact that in the whole of 2011, President Obama made only one speech on energy and made only a single reference to climate change in it.[31]

Notes and references

[1]   By December 2013, they had managed to sell only 116,000 all-electric cars and were nowhere near the 1.5 million mark they had set for 2016. See Nikki Gordon-Bloomfield, 'Ghosn Admits Renault-Nissan

Won't Have Sold 1.5 Million EVs by 2016', Transport Evolved (14 November 2013), available at http://transportevolved.com/ (accessed on 11 July 2017).

2 See http://www.theguardian.com/environment/2009/mar/12/copenhagen-summary (accessed on 11 July 2017).

3 Thomas McCraw, *Prophet of Innovation: Joseph Schumpeter and Creative Destruction.* Cambridge, MA: The Belknap Press, Harvard University (2007).

4 Stern Review, 'Part VI: International Collective Action', in The Stern Review Report, 458.

5 Robert Rapier, 'Cellulosic Ethanol vs. Biomass Gasification', posted on 'The Drumbeat' at website of Oil Drum: Discussion about Energy and Our Future (26 October 2006) for a succinct description of three major hurdles to commercialisation, available at http://www.theoildrum.com/story/2006/10/22/211321/89 (accessed on 11 July 2017).

6 Michael O'Neal, 'Scientists Seek Cheap, Plentiful, Energy Alternatives'.

7 See http://www.businessgreen.com/bg/news/1806738/greenpeace-uncovers-astroturf-campaign-challenge-us-climate (accessed on 11 July 2017).

8 This estimate differs from the one given by Anne Mulern in a *New York Times* report cited below. The difference is small if the data Greenpeace was citing referred to the first quarter of the year.

9 Stephanie Condon, 'Al Gore Urges California to Vote No on Climate Initiative', *CBS NEWS* (20 October 2010).

10 Thomas B. Reed's biography can be found at http://www.woodgas.com/tombio.htm (accessed on 11 July 2017).

11 Allen L. Hammond, 'Methanol at MIT: Industry Influence Charged in Project Cancellation', *Science* 190, (1975, November): 761.

12 Ibid.

13 Ibid.

14 Alastair Gray and Pilita Clark, 'Insurers to Cover Withdrawal of Solar and Wind Subsidies', Financial Times (6 July 2014), available at http://www.ft.com/intl/cms/s/0/badcbca0-0386-11e4-817f-00144feab7de.html (accessed on 11 July 2017).

15 Jason Deign, 'Gemasolar at 2: Beauty in a Beast of a Market', *CSP Today* (30 December 2013), available at http://social.csptoday.com/

markets/gemasolar-2-beauty-beast-market (accessed on 17 April 2014).

16   Ibid.

17   See http://www.eia.gov/todayinenergy/detail.php?id=29192 (accessed on 26 January 2017).

18   Nathan Vanderklippe, 'New Oil Fever Sets Stage for Global Price Pressure; Russia, China, Europe and Saudi firms Rushing to Try Techniques That Took North America By Storm, Energy Expert Says' (Canada: The Globe and Mail, 25 April 2012).

19   Edward Luce, 'Look Away Greens; America is entering a New Age of Plenty', *The Financial Times* (21 November 2011).

20   US Energy Information Administration, 'World Shale Gas Resources: An Initial Assessment of 14 Regions Outside the United States' (5 April 2011), available at http://www.eia.gov/analysis/studies/world-shalegas/ (accessed on 11 July 2017).

21   Edward Luce, 'Look Away Greens'.

22   Daniel Yergin, chairman of energy research firm Information Handling Services Cambridge Energy Research Associates (IHS CERA) believes oil and gas prices will feel the pressure within the next four or five years and, at the very least, stop rising. Thomas Homer-Dixon of the Balsillie School of International Affairs has pointed out while it may depress gas prices, it will at most slow down the rise in oil prices. New shale gas production will augment enormous global conventional supplies of gas. In stark contrast, each year the daily output of the world's mature conventional oil fields drops by 4 million barrels. In five years, this decline is equivalent to twice Saudi Arabia's output—a huge gap that shale oil would not be able to fill. What is more, the output from individual shale oil wells drops much faster than that from conventional wells—often by 80 per cent to 90 per cent in the first two years. So fracking companies must keep drilling at an extraordinary rate to maintain, let alone increase, production. As they exhaust shale oil 'sweet spots' and move to less productive zones, they will use steadily more energy and water to do their job, which will mean steadily more costly, not cheaper, oil. See Nathan Vanderklippe, 'New Oil Fever Sets Stage for Global Price Pressure' and Horner-Dixon in The National Post, aka The Financial Post. National edition, 7 April 2012.

[23]   Anne C. Mulkern, 'Lobbying Cash Paved Climate Bill's Road to the House Floor', *The New York Times* (26 June 2009).

[24]   Martin Wolf, 'Prepare for a Golden Age of Gas' (London: *Financial Times*, 22 February 2012).

[25]   *Scientific American*, 'Ohio Earthquake Likely Caused by Fracking Wastewater' (4 January 2012), available at http://www.scientificamerican.com/article.cfm?id=ohio-earthquake-likely-caused-by-fracking (accessed on 11 July 2017).

[26]   Richard S. Lindzen, 'Global Warming: The Origin and Nature of the Alleged Scientific Consensus', *The Cato Institute* 15, no. 2 (1992, Spring), 87–98.

[27]   Justin Gillis, 'Clouds' Effect on Climate Change is Last Bastion for Dissenters', *The New York Times* (1 May 2012).

[28]   IPCC's fourth Assessment Synthesis report.

[29]   Justin Gillis, 'Clouds' Effect on Climate Change Is Last Bastion for Dissenters'.

[30]   How this happened is described in detail by James Hansen, who was a member of the Task Force. See James Hansen, *Storms of My Grandchildren*.

[31]   Martin Wolf, 'Prepare for a Golden Age of Gas'.

Chapter Ten
Where the World Went Wrong

We have wasted 20 precious years that may well cost human-
ity its future. Where did the world go wrong? The answer is,
at the very beginning.

The way the market actually works, as distinct from the way it is
supposed to work, explains why the governments of the industrialised countries wasted 20 years after James Hansen's unequivocal warning in 1988, and why the commitments they made at
the Paris climate change summit remain utterly inadequate. But
why did the governments of the industrialised countries allow the
market to have its way?

It is not as if they did not know that a threat existed. The
growing doubt about the feasibility of continued exponential
growth on a finite planet that was voiced by the Club of Rome
had led to the first UN Conference on the Human Environment
at Stockholm in 1972. Although the focus of the conference was
upon the threat to the natural and human environment from rapid
industrialisation and the population explosion in the low income
countries, it also issued an unequivocal warning that '… the release
of heat, in such quantities or concentrations as to exceed the capacity of the environment to render them harmless, must be halted in

order to ensure that serious or irreversible damage is not inflicted upon eco-systems'.[1]

The Stockholm Declaration had a profound impact upon peoples' awareness of environmental degradation, in particular the harm that acid rain was causing in the cool temperate regions, but its warning against global warming, which had only just been confirmed by the Mauna Loa Observatory's findings, and the now-famous hockey stick diagram, was barely noticed. In fact, the world had just gone through a 30-year cooling period, so the preoccupation of many scientists after a succession of bitterly cold winters in Europe and the United States was with whether the earth might not be on the threshold of another ice age.[2]

But, as the abrupt termination of Reed and Lerner's experiments with methanol at MIT in 1973 showed, the market had begun to put brakes on the development of alternative energy technologies long before governments became aware of the threat that global warming posed.

It's power to do so was reflected by the scepticism with which governments received the message from Stockholm: Instead of seeking to learn more about the risk the world ran if it continued to grow warmer, they raised the bar of proof that human action was indeed what was causing the unprecedented rise in CO_2 levels and questioned the adequacy of the data and the reliability of its interpretation. This put the climate science community on the defensive, where it has remained ever since.[3]

Among scientists, concern about the possible consequences of rapid global warming only hardened in the 1970s when paleo-climatologists found a way to extract very long core samples of ice from first the Greenland, and then the Antarctic, ice sheets and determine the earth's climate from the composition of the air trapped in the successive layers of snow (compacted into ice) of which these were formed. By the early 1980s, these results, and the Club of Rome's dire warning that mankind's demands could not continue to expand without limit on a finite planet, had prompted UN Secretary General Perez de Cuellar to set up a World Commission on Environment and Development (WCED) under former prime minister of Norway Gro Harlem Brundtland.

In the run-up to every historic event, there is at least one opportunity when a different insight, recommendation or decision could have completely altered the outcome. In the struggle to avert climate change, that opportunity was the formulation of the report of the Brundtland report. Unfortunately, it was this report, of whose energy panel I was the member for renewable energy, that put the world on the wrong track.

The commission began its work in 1984, held annual meetings of its various panels in 1985, 1986 and 1987, and completed its report in 1987. Before presenting it to the UN General Assembly, Mrs Bruntland visited more than a score of countries to get a feedback on its contents. The Commission's report, titled 'Our Common Future', received unprecedented attention in the capitals of the world and became the basic document for the 'Earth Summit' at Rio in 1992, the first concerted attempt to reconcile growth with long-run sustainability. It was in the way that this commission's findings were moulded that the influence of the market became fully visible. For the WCED report added nothing to what had already been reported at Stockholm. The 'six threats' it identified—population growth, lack of food security, growing fragility of the ecosystem and disappearance of species, an excessive dependence on fossil fuels for energy, indiscriminate industrialisation and haphazard urbanisation—were the very ones that the Club of Rome and the Stockholm Conference had identified. Stockholm had limited itself to taking stock of the threat posed by economic development to the environment. It fell to the WCED to propose a programme of action.

But the WCED dodged the challenge. It listed six 'strategic imperatives'—'Ensuring a Sustainable Level of Population; Meeting Essential Human Needs; Changing the quality of Growth; Conserving and Enhancing the Resource Base; Reorienting

Technology and Managing Risk, and merging environment and economics in decision making'—but did not make one concrete suggestion on how they could be met. It did not even tell the world where to start, for that would have forced it to identify the principal cause of the degradation of the environment. There were, after all, only three candidates: rising population, the hunger for economic growth and the evermore frantic extraction of fossil fuels from the earth. Since telling the poor how many children they could have, and asking them to stop trying to catch up with the West was well beyond its terms of reference, it would have had to recommend a rapid replacement of fossilised with renewable energy. But the WCED did no such thing and chose, instead, to forge a diplomatic consensus that was utterly empty of content.

This was not because its experts did not highlight the central role of fossil fuels in degrading the environment, and therefore of the imperative need to develop alternatives. On the contrary, in its report to the Commission, the Energy Panel of the Commission highlighted each and every finding that was emphasised by the IPCC in its fourth assessment report 20 years later. It warned the Commission's members that the consumption of fossil fuels had risen thirty-fold in the 20th century (1900–1985). What was more alarming, the rise had been exponential, for three quarters of this increase had taken place in just 15 years between 1950 and 1985. *If this trend continued unchecked, the level of CO_2 in the atmosphere would double from the pre-industrial level of 280 ppm to 560 ppm by the middle of the 21st century.*

The energy panel also noted that 'Current modelling studies and experiments show a rise in globally averaged surface temperatures, for an effective CO_2 doubling, of somewhere between 1.5°C and 4.5°C, with the warming becoming more pronounced at higher latitudes during winter than at the equator'. 'An important concern', it went on, 'is that a global temperature rise of 1.5–4.5°C, with perhaps a two to three times greater warming at the poles, would lead to a sea level rise of 25–140 centimetres'. The panel then gave a prophetic warning:

There is no way to 'prove' that any of this will happen until it actually occurs. The key question is: How much certainty should governments require before agreeing to take action? If they wait until significant climate change is demonstrated, it may be too late for any countermeasures to be effective against the inertia by then stored in this massive global system. The very long time lags involved in negotiating international agreement on complex issues involving all nations have led some experts to conclude that it is already late. Given the complexities and uncertainties surrounding the issue, it is urgent that the process start now.

But although they devoted an entire chapter of their final report to energy, the Commissioners still preferred to duck the issue of how to cope with the threat that the energy panel had highlighted. Instead they adopted a 'consensus' approach and literally buried this stark warning in a host of other concerns. Only 1 of the 54 paragraphs of its much-hyped call to action, titled 'Our Threatened Future', mentioned global warming. And even that paragraph described the 'greenhouse effect' only as 'one (of several) threat(s) to life support systems (that) spring directly from increased resource use', and put it on a par with the threat from 'industrial air pollution' and 'acid rain'.

The WCED *should* have highlighted what was apparent even to laymen in the 1980s that all of these threats to the environment stemmed from one, and only one, source: the ever-growing use of fossil fuels. Therefore, these were not separate problems but one problem with several manifestations. It *should* have urged the member nations of the UN to tackle these problems at their common root by developing alternative sources of energy to replace fossil fuels. It *should* have reiterated the warning of the Club of Rome that fossil fuels would not last forever. The world had already received two oil price shocks in the 1970s, and no one was under any illusion that the fall in price that had taken place after 1981 would be permanent.

The Commission's mandate, in any case, was to think at least two or more decades ahead. During its preparatory meetings, much of our discussion in the energy panel had revolved around the potential for replacing fossil fuels with renewable sources of energy, if not immediately, then by the early 2000s. Much of it had centred on the state of development of biofuels and solar energy. Biomass gasification had made huge strides in the decade since the Government of New Zealand had proposed gasifying a part of the wood and forestry wastes generated by its vast natural forests of *pinus radiata* to produce transport fuels. The US Department of Energy had funded a number of projects to assess the viability of biomass-based transport fuels, and methanol was being tried out as an alternative to gasoline in places as far apart as South Africa and California.

By the mid-1980s, there were a dozen proven processes for gasifying urban solid wastes to produce synthesis gas. Solar thermal power plants, although primitive by today's standards, were being set up on an experimental basis in California, Arizona, Australia and Israel.[4] There was a consensus in the energy panel that while neither was cheap enough to compete with oil and coal in the mid-1980s, their cost was bound to fall just as the price of oil, in particular, was as certain to fall as the price of oil was bound to rise in the coming years as the most easily exploitable reserves got exhausted.

But the commissioners paid scant attention to these developments because the debate that had occurred within the energy panel on the importance of developing alternate energy sources did not get reflected in its report to the Commission. In its prescriptive portion, the report dwelt at length on ways of cleaning up flue gas emissions and improving the efficiency of automobiles and power stations to deliver more usable energy from them, but made only a passing mention of the need to develop renewable energy.[5]

The Commission was therefore denied the information that would have enabled its members to pose the right questions that would have allowed it to consider a future without fossil fuels. These were the state of development of alternate energy technologies, and their current and estimated future costs. The four-pronged strategy that it recommended was therefore a model of procrastination: (a) develop internationally agreed policies for the reduction of the causative (greenhouse) gases; (b) adopt strategies designed to minimise damage and cope with the climate changes and rising sea level; (c) improve the monitoring and assessment of the 'evolving phenomena' and (d) increase research to improve knowledge about the origins, mechanisms and effects of the phenomena. In short, 'mitigation', 'adaptation', 'monitoring and verification', and further 'research' into the causes of climate change.

The WCED report led to the Rio Conference, nicknamed the 'Earth Summit', in 1992. The summit gave birth to the UNFCCC in 1994. The UNFCCC gave birth to the Kyoto protocol in 1996 and the Protocol has given birth to the 22 Conference of Parties (COPs) meetings that have followed. These four issues have remained the cornerstones of every one of them. In all of them, the 'COPs' struggled mightily to negotiate agreed reductions in CO_2 emissions. They did so in much the same way as they had negotiated reductions in tariff barriers under GATT and the WTO. All these efforts yielded precisely nothing. The concentration of CO_2 continued to rise inexorably till it reached 403 ppm in November 2016. As for the WCED's implied admonition that more research was needed to arrive at concrete findings on the danger that excessive CO_2 releases posed to the environment, all that the research carried of 25,000 scientists, collated in five IPCC reports published between 1990 and 2014, has done is to confirm the accuracy of the data that the Energy Panel had submitted to the Commissioners in the mid-1980s.

In 1987, when the threat from global warming was just beginning to be understood, the commissioners' demand for more research was understandable. But by not endorsing the scientists' warnings, it opened the gate for sceptics to raise the bar of proof higher and higher, until they turned science itself into a tool for procrastination. Thus has the market economy endangered the future of mankind to ensure the profit for the few.

Notes and references

1   This is contained in Principle 6 of the Declaration.

2   Real Climate, 'The Global Cooling Myth', available at http://www. realclimate.org/index.php/archives/2005/01/the-global-cooling-myth/ (accessed on 11 July 2017).

3   The 1975 National Academy of Sciences/National Research Council Report concluded that scientists did not know enough as yet to make accurate predictions and urged the US government to embark on a focused programme to collect more information. It did, however, recommend that an international paleo-climatic data network. See http://www.wmconnolley.org.uk/sci/iceage/nas-1975.html (accessed on 11 July 2017).

4   I was a member of the WCED's energy panel. This was what the representatives of the international corporations, who had besieged the WCED's secretariat in Geneva from the day it was set up, had wanted all along.

5   In a telephonic conversation with the author (who was a member of the Energy Panel), the Canadian secretary general of the WCED conceded that this was what the representatives of the international corporations, who had besieged the WCED's secretariat in Geneva from the day it was set up, had wanted all along.

Chapter Eleven

Backlash:
Attack on Climate Science

The failure of the Copenhagen summit caused a huge back-lash in the environmentalists' community. This was taken advantage of by the Business As Usual lobby to launch a final attack on the credibility of climate science and its scientists.

Copenhagen was where the world finally realised that it had been on the wrong track all along. After two years of hectic prepara-tions, heralded by unprecedented media publicity, the Climate Change Conference in December 2009 proved a fiasco. It did not frame a common strategy to combat global warming, did not even draw up the skeleton of a plan and did not make a single commit-ment that would even remotely slow down the annual rise in CO_2 and other GHG emissions, much less reduce them. Apart from Japan and some members of the EU, not one of the larger coun-tries was prepared to discuss a binding, verifiable commitment to cut GHG emissions. Nor were they willing to make more than token contributions to a climate fund that could be used to subsi-dise the adoption of new energy technologies. Its only 'success', if one can call it that, was to bury the Kyoto Protocol by making all future commitments to cut CO_2 emissions voluntary. This freed the industrialised countries from the only commitment that they had ever made. It was humanity's biggest collective failure since the failure of the League of Nations to avert the Second World War.

The fiasco at Copenhagen forced people to reassess the actions of their governments through disillusioned eyes. There could be only two explanations of why governments preferred to bicker endlessly over secondary issues such as apportioning blame and responsibility for rising CO_2 levels, sharing of costs of mitigation and choosing away to verify CO_2 reduction claims, when the survival of humanity (and most other life forms) was possibly at stake: they either did not themselves believe in the threat and were using it to meet other goals or were willing to barter the survival of the human race for short-term personal gain. Since the second was virtually unthinkable, most people concluded that it had to be the first. The immediate impact of the failure at Copenhagen was, therefore, to discredit the science of climate change itself. Not surprisingly, the ranks of the climate change sceptics swelled overnight. An American Gallup poll published in March 2010 showed that in the United States, concern over global warming had declined sharply, for 48 per cent of Americans believed that it was exaggerated, against 35 per cent two years earlier.[1]

The attack on climate scientists

On balance, it was the scientists, and not the governments, who bore the brunt of peoples' anger. Two relatively minor disclosures in the weeks before the conference provided them with targets to attack. In November 2009, unidentified computer hackers down-loaded more than a thousand messages exchanged by scientists studying climate change from the computers of the University of East Anglia Climate Research Unit (CRU). Selected mails, which purported to show how the scientists were covering up or manip-ulating data, were then uploaded almost simultaneously to servers in countries as far apart as Turkey, Russia and Saudi Arabia. These revealed that the famous 'hockey stick-shaped' graph, which had first shown the world how CO_2 emissions had shot relentlessly upward after 1958, was a fabrication. To construct it, scientists had merged data obtained from two sources but had ignored a third source, the thickness of tree rings, which suggested that there had been a decline in temperature after 1950.

The hackers targeted two scientists in particular, Professor Phil Jones, the Head of the East Anglia's CRU and Professor Michael Mann of Pennsylvania State University in the United States and accused them of using a 'trick' to arrive at a predetermined conclusion.

Himalayan glaciers not shrinking

Close on its heels came a second 'revelation'. An Indian review of the behaviour of Himalayan glaciers over the past half-century had shown that contrary to a finding contained in the IPCC's fourth assessment report, they are not retreating and are not, therefore, likely to disappear by 2035 or 2040. The IPCC, according to the report, had based its warning upon a remark made by an Indian glaciologist to a non-peer reviewed journal, the *New Scientist*.

The East Anglia story was broken by a columnist for *The Telegraph* (of London), James Delingpole, a leading climate sceptic (whose claim on his blog is that 'he is always right'), on 20 November, barely a fortnight before the Copenhagen Summit. The story attracted only limited attention in the mainline media when it first appeared. But in the disillusionment that followed the failure of the summit, it acquired a vast new significance and started being referred to as 'Climategate'. After the failure at Copenhagen, the word was reportedly used nine million times in the first week.

Delingpole also dismissed the IPCC's warning about the disappearance of the Himalayan glaciers by claiming, derisively, that

in the course of a garbled phone conversation a scientist accidentally invents a problem that doesn't exist. This gets reported as if gospel in an influential 'Warmist' science magazine and repeated by a 'Warmist' NGO, before being lent the full authority of the IPCC's fourth assessment report which, as we know, can't be wrong because it is vetted by around 2,500 scientists.[2]

'The Resilient Earth', a leading climate-sceptic website went a long way further and claimed that the IPCC's warning was based on 'a third hand speculation from a little known scientist'.[3]

The damage to the IPCCs credentials was vastly increased when a variety of scientists from all quarters of the globe, including several whose submissions had not been included in the IPCCs assessment report, joined in the attack. 'The Resilient Earth' reported a paper by Jeffrey S. Kargel, a glaciologist at the University of Arizona, which cited evidence gathered by a team from the University of Nebraska that 'many glaciers in the Karakoram Mountains, on the borders of Pakistan and India, have stabilised or undergone an aggressive advance'. A Canadian glaciologist echoed these findings, saying that in the Mount K-2 region of Karakoram Range of the Himalayas, he had observed five glaciers that were advancing and only one in retreat. John Shroder, also of the University of Nebraska, concluded that the IPCC's Himalayan assessment had got it 'horribly wrong'.[4]

The tabloid press was only a shade less restrained. David Rose of London's *Daily Mail* wrote:

> The scientist [Dr Murari Lal] behind the bogus claim in a Nobel Prize-winning UN report that Himalayan glaciers will have melted by 2035 last night admitted it was included purely to put political pressure on world leaders.
>
> Lal also said he was well aware that the statement, in the 2007 report by the IPCC, did not rest on peer-reviewed scientific research.
>
> In an interview with *The Mail* on Sunday, Lal, who was the co-coordinating lead author of the report's chapter on Asia, said: It related to several countries in this region and their water sources. We thought that if we can highlight it, it will impact policy-makers and politicians and encourage them to

take some concrete action. It had importance for the region, so we thought we should put it in.[5]

This admission proved grist to the mill of the climate sceptics.

Just when the heat generated by these accusations was at its peak, there came a third attack on the credentials and integrity of climate scientists and of the IPCC itself. On 31 January 2010, *The Sunday Telegraph* claimed that the IPCC's warning that a small fall in the rainfall in the Amazon, caused by global warming, could wipe out 40 per cent of the rainforest and replace it with Savannah, was 'based on an unsubstantiated claim by green campaigners who had little scientific expertise'.

'The source for its claim', wrote the reporter, Jonathan Leake, was

> [A] report from the World Wildlife Fund that was authored by two green activists [who] had based their 'research' on a study published in *Nature*, the science journal, which did not assess rainfall but in fact looked at the impact on the forest of human activity such as logging and burning.... This [was] the third time in as many weeks that serious doubts ha[d] been raised over the IPCC's conclusions on climate change.[6]

Ad hominem attacks

The attacks did not remain confined to the data, but were quickly extended to the motives and morals of the scientists. In the weeks after Copenhagen Conference, R.K. Pachauri, the chairman of the IPCC, became the climate change sceptics' target of choice. In an article published in *The Sunday Telegraph*, London, in December 2009, titled 'Questions Over Business Deals of UN Climate Change Guru Dr Rajendra Pachauri', two journalists Christopher Booker and Richard North alleged that Pachauri had been 'making a fortune from his links with "carbon trading" companies' and claimed that the money he had made while working for other organisations 'must run into millions of dollars'.[7] Anyone familiar with Booker's writings would know that he is a journalist who revels in taking contrarian positions, often at the expense of truth and objectivity. A co-founder of *Private Eye* in 1961, Booker has, among other things, insisted that neither passive smoking nor asbestos expose human beings to the risk of cancer, and that Darwin's theory of evolution has been rendered obsolete by the theory of 'intelligent design'. But *The Sunday Telegraph* is a widely read newspaper and virtually a bible for the right wing of the conservative party in Britain.

An array of British policy makers, from the former conservative Chancellor of the Exchequer Nigel (Lord) Lawson, to the head of the Grantham Research Institute on Climate Change and the Environment at the London School of Economics, therefore, urged that there should be a 'rigorous enquiry' into the content of the East Anglia e-mails. In a remark that brought little credit to his organisation, John Sauven, the director of Greenpeace UK, baldly demanded that R.K. Pachauri be replaced as head of the IPCC.[8] Not a single government or leader came to the defence of the scientific community. Ed Miliband, the then British minister for the environment, hastily admitted that 'mistakes have been made'. Gordon Brown chose to remain silent. So did Obama across the Atlantic.

Left on their own, the scientists were forced to give ground. Pachauri had dismissed the initial accusations about the Himalayan glaciers as being based on 'voodoo science' and accused the Indian Ministry of Environment, Forest and Climate Change— which had issued the report claiming that the glaciers were not disappearing—of 'arrogance' in rubbishing the IPCC's assessment. But the sustained attack forced the IPCC to issue an apology of sorts: in a joint statement he, the vice-chairperson of the IPCC, and the co-chairs of the IPCC Working Groups reasserted that 'glacier mass losses, including in the Himalayas, are likely to accelerate throughout the 21st century, reducing water availability in regions supplied by melt-water from mountain glaciers and snow packs', but conceded that 'in drafting the paragraph in question, the clear and well-established standards of evidence required by the IPCC procedures were not applied properly'.[9]

An idea of the pressure that their hasty abandonment by the governments of the principal polluting countries put upon climate scientists can be had from a hasty disclaimer added by the World Wildlife Society to a 2005 report on the state of Nepal glaciers. Under the rubric of 'Correction', it admitted that its endorsement of a statement in a 1999 report of the Working Group on Himalayan Glaciology (WGHG) of the International Commission on Snow and Ice (ICSI) that 'glaciers in the Himalayas are receding faster than in any other part of the world and, if the present rate continues, the likelihood of them disappearing by the year 2035 is very high' was used in good faith but 'it is now clear that this was erroneous and should be disregarded'.[10] Thus, by the end of February 2010, only eight weeks after the end of the Copenhagen Conference, the global media had persuaded the public to conclude that global warming was a hoax, perpetrated by a cabal of scientists hungry for power, recognition and pelf.

A barrel of lies

The year-long investigations that followed the allegations showed that both the accusations were unfounded. But in contrast to the furore the allegations had created, few outside the climate science community came to know of the vindication. When the storm broke, the CRU at East Anglia had explained that it had ignored the Woodring data because it had directly observed temperature data from every part of the planet for the period after 1950 which showed that there had been a continuous rise in air temperature in the ensuing decades. It had, therefore, treated the Woodring data as an anomaly that required a locally specific explanation. Responsible scientists from a host of other fields had supported its action, pointing out that when data from a variety of sources all told the same story, the conclusions drawn from them was not invalidated by anomalous data only from a single set of observations.

In an attempt to clear the air and restore the credibility of scientific research, Pennsylvania State University set up a formal review board to examine the charges that had been made against Professor Michael Mann, the principal author of the hockey stick diagram. After months of painstaking inquiry, the board concluded that the so-called 'trick' that Mann was accused of playing to construct the 'hockey stick' graph of global warming was a commonly used statistical method to bring two or more different sets of data together to tell a continuous story. It said that it had arrived at this conclusion after the use of this technique had been reviewed by 'a broad array of peers in the field'. Responsible scientists working in other fields concurred with the board's finding.[11]

On 30 March 2010, six weeks after the Pennsylvania State University review board exonerated Mann, the Science and Technology Select Committee of the UK House of Commons also dismissed the charge that the two East Anglia scientists had conspired to present only data that pointed to a predetermined conclusion. A further review led by Sir Alastair Muir Russell, chairman of the Judicial Appointments Board for Scotland, found no fault with the 'rigour and honesty as scientists' of Jones and his colleagues.

As for the IPCC's 'error' about the Himalayan glaciers which spooked Sauven, Greenpeace UK, into abandoning IPCC's Chairman Pachauri in his hour of need, it soon became apparent that neither Sauven nor anyone in Greenpeace had actually read the Indian Ministry of Environment, Forest and Climate Change's report. V.K. Raina, its author, had reported that the average retreat of the Himalayan glaciers, which had been about five metres a year from the 1840s when records began to be kept, had accelerated 'manifold' between the 1950s and 1990s, but had slowed down after the mid-1990s till, in some of the country's biggest and best-known glaciers such as Gangotri and Siachen, 'it has practically come down to stand still during the period 2007–09'. Raina also found a substantial variation in the behaviour of often-adjoining glaciers. He, therefore, concluded that 'It is premature to make a statement that glaciers in the Himalayas are retreating abnormally because of the global warming'.[12]

Raina's caution was understandable, but his conclusion was, to say the least, hasty because it related to only one of several methods of gauging the behaviour of glaciers. Raina had measured the retreat of these glaciers solely from the position of their snouts. But in his report he had made it clear that the retreat of the snout is only one of three measures of glacier change. The other two are its 'mass balance', that is, the volume of ice in it, and the rate at which the ice is melting. The most accurate measure of the shrinkage of a glacier is its loss of mass. Raina's study contained unambiguous data which showed that the Himalayan glaciers had not stopped losing mass.[13] The rate of loss had been on an average higher in the north-western Himalayas and lower further east. It had also varied considerably from one glacier to the next, but the overall loss had been substantial and had not significantly slowed down.

Proof of this came from evidence collected by the Indian Space Applications Centre (ISAC) at Ahmedabad. Using remote-sensing technologies to study 466 glaciers in the central Himalayas over 42 years from 1962 to 2004, the ISAC concluded that these glaciers had shrunk by 21 per cent in area and 30.8 per cent in volume. The ISAC's data also showed that small glaciers were losing mass at a much faster rate than larger ones, for those with an area of less than 1 km^2 had lost 38 per cent of their area and about 55 per cent of their mass during this period. A separate tabulation for 26 larger glaciers in Sikkim showed that they had retreated four times faster between 2000 and 2005, than between 1976 and 1978. Only 1 of the 26 showed an advance, while a second showed both a rapid retreat and a slower advance in different periods.

Raina also did not consider the possibility that, like the tree rings, the stabilisation of the snout of a glacier could also be an anomaly caused by local influences. Had he done so he would have seen that his data are open to at least one other explanation. This is the deposit of dust and black carbon (collectively called aerosols) upon the ice. In a different part of the report, Raina cited laboratory studies that had shown that till an accumulation of 400 g/m², the deposit of aerosols raises the rate of melting. Between 400 and 600 g, it becomes neutral, but when it exceeds 600 g it acts as a shield against the sun and slows down melting. But he did not link the stabilisation of the snouts to the rising level of aerosol deposits in the lower reaches of the Himalayas caused by overpopulation, overgrazing and deforestation.

Had he done so he might have found an explanation for why the snout of the Gangotri glacier has virtually stopped receding while the Siachen glacier, India's largest, advances and retreats in a regular 25-year cycle. The Gangotri lies in one of the most densely populated parts of the Himalayas. The area around Siachen, by contrast, is almost uninhabited.

Aerosol deposits can also explain another anomaly that Raina recorded but did not explain: Why there has been a noticeable narrowing of the middle part of several glaciers of the glaciers he has studied, including the Gangotri, so much so that some glaciers have split into two sections with a barer patch in between? The most likely explanation is that the snout has been preserved lower down, where aerosol deposits exceed 600 g/m² and have narrowed further up where the deposits fall below 200 g/m².

There is, clearly, a need for continued research on the impact of aerosols on the behaviour of glaciers, but new findings are not likely to alter basic fact that the behaviour of glaciers in the Himalayas has been little different from the behaviour of glaciers in the rest of the world. While there have been local variations in all areas, the average for glaciers across the world has been remarkably uniform—the vast majority are in retreat. A recent World Bank study showed that Bolivia's famed Chacaltaya glacier had lost 80 per cent of its surface area since 1982, and Peru's glaciers had lost more than one-fifth of their mass in the past 35 years, reducing water flow to the country's coastal region, home to 60 per cent of Peru's population, by 12 per cent.[14]

Pachauri exonerated

As for Pachauri's moral culpability, the IPCC chief had first rubbished the allegations, but the rising clamour for his head forced him to take Booker's and North's attack seriously. He, therefore, decided to submit both his personal finances and the accounts of The Energy and Resources Institute (TERI), which he then headed in India, to professional scrutiny. The report of the auditors, published on 26 August 2010 in *The Guardian*, showed that the consultancy fees that Pachauri earned went not to him but to TERI and that, contrary to a common practice in research and teaching institutions worldwide, none of it was retained by him. Pachauri received only his annual salary of £45,000 a year. His additional income from outside earnings over the 20 months, reviewed by Klynveld Peat Marwick Goerdeler (KPMG), amounted to £2,174. In sharp contrast to the condemnation that had been heaped upon him, Pachauri's exoneration got barely a mention in the international media.[15]

Greenwash

In truth, governments, especially of the industrialised countries, have betrayed their own people and have done so to fatten the coffers of the oil lobby, the armaments industries and the military. There is an abundance of evidence to support this conclusion: the rich nations' headlong rush into 'fracking' and extraction of 'tight oil'; their refusal to declare the highly Arctic ocean a common heritage of mankind (like the Antarctica) and determination to extract oil and REs from Greenland's continental shelf[16] and their determination to force open the north-west passage through the Arctic and keep it open for most, if not all, of the year even though this will expose still more of its dark, heat-absorbing, waters to the rays of the sun.

Time and again, in the years since the Second World War, they have chosen to throw their hands in with the oil and gas industries. Whenever these industries have faced a threat, whether from a new technology or a foreign government, their governments have intervened on their behalf to delay, if not suppress, the development of an alternative energy technology or topple the offending government. Today, the moving force behind the new gold rush in the Arctic is the immensely powerful global oil and gas lobby. Just as the coal-based power industry lobbied vigorously with the WCED secretariat in Geneva in 1987, in favour of supercritical turbine technology and expensive flue gas cleaning systems that would deliver clean energy from coal,[17] the oil industry lobbied equally vigorously with the US Congress against the Waxman–Markey Bill (designed to cap CO_2 emissions in the United States) in 2009.

For the past quarter century, the governments of all the major economies have been speaking through both sides of their mouths. In 1991, the eight nations of the Arctic Council were wholeheartedly supporting the push to exploit the resources of region even as they were signing an Arctic Environmental Protection Strategy (1991) that committed them to 'promoting cooperation, coordination, and interaction among the Arctic States, with the involvement of the Arctic Indigenous communities and other Arctic inhabitants on issues such as sustainable development and environmental protection'.

At Copenhagen while the OECD countries staunchly refused to create a special fund to subsidise the introduction of alternative energy technologies, they and other energy-importing countries were subsidising fossil energy consumers to the tune of $135 billion dollars a year.[18] Their recent actions have made it crystal clear that they will continue to do so even when they know that every new shred of evidence that has been collected since 1987 when the WCED tabled its report has only confirmed what scientists already knew then—that only a drastic shift out of fossil fuels can save the world we know.

No individual or group has claimed credit for hacking into East Anglia's computers, and no one has identified the hackers. No one knows how the information was disseminated so rapidly in the days that followed. But all of this required a high degree of organisation and sophistication. The release of the information occurred only a few weeks before the start of the Copenhagen Conference and was timed to coincide with a spate of new books that sought, one final time, to discredit climate science before the world formally adopted CO_2 reduction targets that would force a shift out of fossil fuels.[19] Finally, to make doubly sure that the attack on climate science succeeded, those behind it chose to attack the moral rectitude of the scientists. All this happened within a time span of only eight weeks could have been a coincidence, but when so much is at stake it would be naïve to believe that this is all it was. On the contrary, the entire operation reeked of power, organisation and money—lots of money. James Hansen's forthright accusation that the United States and other governments are lying when they pretend that the challenge is only to slow down the accumulation of CO_2 when they know fully well that it is to bring *down* its concentration in the atmosphere at least to the 1950 level, and his contemptuous description of the plethora of 'green' initiatives as 'greenwash' is therefore easy to understand.

Notes and references

[1] Reported in the *Washington Independent* and other newspapers, 11 March 2010.

[2] James Delingpole, 'Syed Hasnain, R.K. Pachauri and the Mystery of the Non-disappearing Glaciers', *The Daily Telegraph* (18 January 2010), available at http://blogs.telegraph.co.uk/news/jamesdelingpole/100022694/syed-hasnain-rk-pachauri-and-the-mystery-of-the-non-disappearing-glaciers/ (accessed on 11 July 2017).

[3] See http://www.theresilientearth.com/?q=content/himalayan-glaciers-not-melting (accessed on 1 April 2010).

4 Ibid.

5 David Rose, 'Glacier Scientist: I Knew Data Hadn't Been Verified', *The Daily Mail* (24 January 2010).

6 Jonathan Leake, 'UN Climate Panel Shamed by Bogus Rainforest Claim', *The Sunday Telegraph* (31 January 2010).

7 The Telegraph, 'Christopher Booker's Notebook', *The Sunday Telegraph* (7 August 2005).

8 Ben Webster, 'IPCC Chief Rajendra Pachauri Under Pressure to Go over Glacier Error' (London: *The Sunday Times*, 4 February 2010).

9 See http://blogs.nature.com/climatefeedback/2010/01/ipcc_apolo-gises_for_himalayan.html (accessed on 1 April 2010).

10 WWF, 'An Overview of Glaciers, Glacier Retreat, and Subsequent Impacts in Nepal, India and China' (WWF Nepal Program, 2005, March).

11 Henry C. Foley, Alan W. Scaroni and Candice A. Yekel, 'RA-10 Inquiry Report: Concerning the Allegations of Research Misconduct Against Dr. Michael E. Mann, Department of Meteorology, College of Earth and Mineral Sciences, The Pennsylvania State University' (The Pennsylvania State University, 3 February 2010), available at http://www.research.psu.edu/orp/Findings_Mann_Inquiry.pdf (accessed on 1 April 2010).

12 V.K. Raina, 'Himalayan Glaciers: A State-of-art Review of Glacial' (Ministry of Environment Discussion Paper, 2009).

13 Ibid. The executive summary states without qualification 'All the glaciers under observation, during the last three decades of 20th century have shown cumulative negative mass balance'.

14 Reported by Carolyn Kormann in Environment 360, 'Retreat of Andean Glaciers Foretells Global Water Woes' (9 April 2009), available at http://e360.yale.edu/content/feature.msp?id=2139 (accessed on 1 April 2010).

15 George Monbiot, 'The Smearing of an Innocent Man', *The Guardian* (26 August 2010).

16 The US Geological Survey has assessed that there are approximately 90 billion barrels of oil, 1,669 trillion cubic feet of gas and 44 billion barrels of natural gas liquids (NGLs) that remain to be found in the Arctic. Of the total 412 billion barrels of oil equivalent (boe), approximately 84 per cent is expected to be found offshore, and about two-thirds (67 per cent) of the total was natural gas. It is estimated that

there are 20 billion barrels of oil and 20,000 tonnes of REs buried in the Greenland plateau and continental shelf. See http://www.ey.com/Publication/vwLUAssets/Arctic_oil_and_gas/$FILE/Arctic_oil_and_gas.pdf (accessed on 17 July 2014).

The Kvanefeld Limaussac reserve, Greenland's largest reserve, is estimated to contain 10.3 million tonnes of RE oxides with an average RE content of 1.07 per cent. See http://www.world-nuclear.org/info/Nuclear-Fuel-Cycle/Uranium-Resources/Uranium-From-Rare-Earths-Deposits/ (accessed on 11 July 2017).

[17] This conclusion is based upon the personal experience of the author, who was a member of the Energy Panel of the WCED. He was told this by Jim Wilson, the secretary general to the Commission, in a private conversation when he asked why so little of the Energy Panel's discussion of renewable energy alternatives was reflected in its report to the commissioners.

[18] See http://www.iea.org/media/weowebsite/energysubsidies/ff_subsidies_slides.pdf (accessed on 11 July 2017).

[19] Notable both for the publicity it received and the blunders it made was Australian Geographer Ian Plimer's *Heaven and Earth*.

Part IV

Gateway to a Better World

Chapter Twelve

How the Sun Will Change Our World

What will a post-fossil fuel world look like? This chapter gives some answers, and they are all good. One important conclusion is that the shift of energy base to the sun will benefit the countries of the global South more than those of the North. Income inequalities between the two will narrow.

Looking back at the speed with which extreme weather events have multiplied, it is difficult to remember that until barely a decade ago, the world had been almost completely unaware of the threat that lay nested within the rapid warming of the earth's atmosphere. Today, as a wave of something close to panic has begun to sweep the world, the climate change sceptics are fighting a losing battle. At the Climate Change Summit in Marrakesh, in December 2016, the World Meteorological Organisation, which had been the very first organisation to sound the alarm about global warming in 1969, announced that fully half of the extreme weather events of the past few years could be traced to human-induced global warming. But, more than their frequency, it is their nature and location that has given rise to the current wave of concern, for the most anomalous of these events—the shift of the Jetstream; the smashing of the Polar vortex, above zero temperatures at the North Pole during the winter solstice, and plumes of methane escaping from the seabed—have taken place in or around the single, climatically most sensitive, spot on the planet—the Arctic Ocean.

These events are vindicating the guarded prediction by the IPCC in its Fifth Assessment Report that the rise in the mean atmospheric temperature will accelerate throughout the 21st century, for almost any rate of growth of emissions, and that in the worst case scenario it would be of the order of 3.7°C over the average of 1982–2005, i.e., 4.5°C above pre-industrial times, between 2081 and 2100.[1] Given the uneven distribution of the increase over different parts of the world, this will mean an *up to 11°C rise in temperature in the arctic region.* Shorn of language needed to pass muster with 120 governments, this is a warning that long before the world gets there global warming will be out of control, and humanity's fate will be sealed.

To give our grandchildren, and their grandchildren, a chance of surviving the 22nd and entering the 23rd century, we have not merely to bring our CO_2 emissions down to zero well before the end of this century, but extract large amounts from the air to bring down its concentration to the 312 ppm level of 1950, that Hansen had considered the safe maximum a decade ago.

The technologies described in middle chapters of this book can do this. So rapid is the spread of information today, and so great the accumulation of capital in search of new avenues of investment, that it is technically possible for the world to complete a shift out of fossil fuels in as little as 40 years–the average lifespan of most oil fields, and of modern coal-based power plants today. That will leave 40 more years in this century for the earth's natural carbon sinks to suck up to 800 GT of CO_2 out of the atmosphere. Salvation is therefore still possible but hangs by a hair, and will only occur if the world's governments succeed in harnessing the immense power of global capital to the task, instead of demanding that it reduce emissions without telling it how as they have been doing so far. To do that, these technologies have to be made not only profitable to investors but also free of risk in the initial, pioneering phase of their development. As the data on costs and returns given in the middle chapters shows, all the technologies discussed in them, biomass gasification to produce transport fuels and synthetics, solar thermal power plants to generate power and process heat, and the extraction of hydrogen from water without electricity, are very close to being commercially viable. But the first two have been stymied three times over the past four decades by sudden and erratic price falls in a completely unregulated, and chaotic, global market that no Western government wishes to regulate.

This has to change. What investors need today is not just a carbon tax, although that would undoubtedly speed up the transition, but an international commitment to stabilise the long-term domestic prices of power and transport fuels.

Not only is this not much to ask for, but it would serve the interests of the investors themselves, so why are the global corporations and so many governments bitterly opposed to it? Ideological blinkers aside, the only other explanation is that, like all other life forms, human beings are conditioned to react only to immediate danger. Distant threats, even those we fully understand, do not trigger the instinct for self-preservation.

The threat looks especially remote to the inhabitants of the rich nations, for they inhabit the cool temperate regions of the planet, which will suffer the least from the pole-ward advance of the deserts, the increased frequency of droughts and floods, and the inundation of tens of millions of acres by the sea that global warming will cause. Till the summer of 2015, Western Europe and the southern states of the United States were confident that they had the means to 'repel boarders' and were not afraid to use them. The mass migrations from Libya, the Sahel countries of North Africa, Syria and Iraq have shattered that confidence, but no one has, as yet, realised that these are harbingers of worse things to come. We, therefore, need to find more immediate reasons for shifting out of fossil fuels.

Fortunately there is no dearth of them, and all can be summed up in one short paragraph: The shift from fossil fuels to solar energy in all of its myriad forms will invalidate the Club of Rome's 1972 prediction of a systemic collapse of civilisation by the middle of this century into the distant future by removing the two constraints to growth that are most capable of making it happen—the exhaustion of non-renewable resources and the accumulation of CO_2.

The former will eliminate one of the main causes of war in our time, and consequently of the wreckage of frail economies, the failure of borderline states, the rise of terrorism and the descent of hordes of desperate starving people onto Europe and America by setting off a 'virtuous' circle of growth in the poorest countries of the earth, and make them the drivers of future global industrial growth. For, these are the regions which are most generously gifted with solar radiation, equable climates and an abundance of riotously growing biomass. As the development of the steam engine and the automobile did in the 18th and 19th centuries, the harnessing of solar thermal, biomass and, a little later, hydrogen energy will set off a huge new surge in ancillary technological development. This will lead to the design and manufacture of an entire new generation of machines that will harness the new sources of energy in evermore ingenious ways. And it will do all this while leaving the distribution and consumption structure of society virtually unaffected.

For instance, the entire boiler and turbine sections of power stations will need no redesigning—for central tower solar power stations are already providing the supercritical steam needed by the ultra-supercritical power stations being set up today. The transmission grids will also, therefore, need no alteration and no step-up transformers (that treble the cost of feeding solar PV energy into a national grid). By the same token, there will be no need to replace, or transform the vast network of transport fuel supply that the world has built over the past century. Biofuels can be pumped through same pipes, tanks and petrol pumps that supply gasoline, diesel or ATF today. The social cost of shifting one's energy base will, in both cases, be close to zero. In any case, setting up a new infrastructure to fuel electric cars at petrol pumps, or supply hydrogen to fuel cell powered vehicles will take time, and time is precisely what the world has run out of.

The most profound change that the shift from fossil fuels into solar energy will bring about, however, is one from which all countries, rich and poor, will benefit. This is to re-forge the organic connection between industry and agriculture, and between town and country that was broken when coal replaced wind and water as the main driver of industry, oil and gas replaced the horse as the source of transport energy, and synthetic raw materials progressively replaced cotton, wool, linen, wood and stone as the raw materials for clothing, furnishing, building and transport. This had turned the countryside into an economic backwater.

If biomass becomes the source of transport fuels and petrochemicals, agriculture will recover much of the primacy it has lost in the last two centuries of fossil fuel-based growth. After four decades of gradual de-industrialisation, culminating in a decade of stagnation, shifting the energy base of the global economy will give the industrialised world the stimulus it needs for reviving sustainable economic growth.

An industrial renaissance

As the Gemasolar plant in Spain has already demonstrated, even in the temperate regions solar energy can supply much of the power that is now being produced from coal, nuclear and hydro-power plants. Shifting out of fossil fuels into solar and biomass energy will open up scores of new avenues for profitable investment, from the mass production of components such as solar panels, heliostats and computerised control systems for power stations, to plasma and other designs of gasifiers and the entire array of plant and equipment that is needed to turn the synthesis gas into transport fuels and petrochemicals.

There are also vast amounts of agro-residues in the industrialised countries. Using these as feedstock for producing transport fuels will increase farm revenues sharply and reduce, perhaps even make unnecessary, the subsidies that high income countries are giving to their farmers. This will release much-needed funds for social security and infrastructure maintenance and development.

The potential for industrial rejuvenation can be judged from the fact that in 2013 the energy majors spent $1.73 trillion on extracting and distributing fossil fuels, and that this is expected to rise to $2.55 trillion by 2035. In all, the Paris-based IEA expects $48 trillion to be invested in energy extraction and distribution between 2014 and 2035. If even half this amount is invested in solar power and the production of biofuel via gasification, it will trigger a burst of investment that will end the current recession, even reverse the long-term decline of manufacturing and create millions of new jobs, thereby reducing the growing incidence of chronic unemployment and the progressive immiserisation of the middle class. It will above all give back a secure future to young people and enable them to marry, buy homes and plan families as their parents and grandparents did. Even if much of the actual manufacturing of new products continues to be done in low wage countries, new jobs will have to be created in all the 'non-tradable' sectors of the economy—notably construction and the service industries, education and health.

Even that would only be the beginning of the transformation that the industrialised world has been yearning for. As happened after the development of the steam engine in the early 19th century and with cyberspace in the 1960s, the shift to a new energy base will trigger a spate of breakthroughs that will spread through society like ripples in a pool. Today, the United States is relying on 'fracking' to impart this kind of spurt to design and manufacture in the engineering, chemicals and construction industries. The shift into biofuels described above will do it on a far larger scale, and in every industrialised country, without triggering fresh conflicts over ownership and exploitation rights of the kind that are already looming over the horizon in the case of fracking and extraction of 'tight oil'.

Finally, when the technologies now being developed for using sunlight to produce hydrogen from water mature, coal-based power plants will be able to combine them with some of their flue gas CO_2 to produce synthetic fuels at a fraction of the cost of processes based upon conventional electrolysis. This will allow coal-fired power plants to lower emissions, increase their revenues, extend the lives of their power plants and, through the accumulation of cash reserves, create the capital base that will enable them to switch to solar power at a time of their choosing.

A cleaner world

The most immediate will be an unrecognisably cleaner environment and a sharp improvement in public health, especially in the low income countries. If urban waste becomes the feedstock for producing ATF, and diesel, or other transport fuels, not one speck of it will be left on the streets. If methanol replaces gasoline, vehicle exhausts will consist almost entirely of water vapour and carbon dioxide. There will be no particulates and no sulphurous and nitrous oxide gases. A host of respiratory diseases, to which children and the very poor are particularly prone, will, therefore, become much rarer. If bio-ATF replaces conventional jet fuel, the damage being done to the ozone layer by the relentless build-up of CO_2 will be arrested.

Large amounts of land close to cities, which would otherwise have had to be acquired for landfills, will become available for schools, parks and housing. In cities that are currently incinerating their solid wastes, it will end the discharge of highly toxic dioxins and furans into the air. Gasification, in short, will turn what is a nuisance and a serious health hazard into a valuable resource.

In the cities of the rich nations, it will replace increasingly complicated and inefficient expensive systems of garbage sorting with a single, revenue-earning, method via plasma gasification. For while garbage will still have to be sorted in low income countries to produce refuse derived fuel (RDF), in rich nations where disposal is the main concern, it can be fed directly into the gasifiers without pre-sorting. In addition to producing synthesis gas that can be turned into ATF, it will yield 50 to 100 per cent more process heat for use by industry and homes.

Arresting deforestation and other dangerous changes in land use

A shift to biomass gasification will also eliminate another serious threat to our planet's dwindling forests. This is their replacement by palm oil plantations designed not to meet the need for food but biofuels, specifically biodiesel. Forests have been vanishing for decades to make way for food, cash crops, pasturage and timber for the housing and furniture industries. But the decision of the United States and EU to reduce their consumption of diesel by replacing a part of it with biodiesel obtained from palm oil has opened a new path to their destruction. The way in which the Brazilian government's decision to increase acreage under sugar cane to meet the United States' demand for methanol, and the diversion by US farmers of corn from feeding cattle to producing ethanol, has set off a sequence of changes in land use across Latin America that have accelerated the clearing of the Amazon forest in Brazil has already been described.[2] Exactly the same shift in land use is accelerating the destruction of the world's other great CO_2 sink, the forests of Malaysia and Indonesia.

In the 1970s, Malaysia and Indonesia, then the possessors of forests as vast as the Amazon, began to cut down thousands of hectares every year to grow palm oil for human consumption and became the largest palm oil producers in the world. In August 2006, Malaysia shipped its first consignment of 60,000 tonnes of biodiesel, made from palm oil, to Germany, and Indonesia announced that it would soon follow suit. The EU welcomed this far-sighted initiative because it would save 168,000 tonnes of CO_2 and other GHG emissions. What it chose not to remember was that the palm oil being used to make biodiesel had required, at an earlier date, the cutting down of 12,000 hectares of rainforest, and that the hardwood trees that had stood there had stored an average of 235 tonnes of carbon per hectare. The deforestation had, therefore, released approximately 2.8 million tonnes of carbon, that is, almost 10 million tonnes of CO_2 into the atmosphere. The palm oil trees that were planted in their place would fix at most 2 million tonnes. 'Thus to save one tonnes of CO_2 emissions in Europe, the OECD is overlooking the release of 48 tonnes CO_2 into the atmosphere in Asia'. The Hamburg-based Environmental Protection Encouragement Agency (EPEA) estimated that new palm oil plantations would take 59 years to fix the amount of carbon that was released when the rainforest was cut down. By then, struggle against global warming would itself have become history.[3]

But these figures, and the entreaties of its own environment protection and human rights organisations, have meant nothing to the Malaysian government. Shortly after its initial venture into biodiesel exports, Malaysia launched a 'B5' programme designed to blend 5 per cent of biodiesel with normal diesel for domestic consumption. Under this programme, it produced 249,213 tonnes of palm oil diesel in 2012, of which it blended 113,000 tonnes into the diesel sold in the Malay Peninsula. In 2014, Malaysia decided to extend the B5 programme to the southern part of the country and raise the blending ratio to 10 per cent. To do this, it is increasing the production of palm oil diesel to 500,000 tonnes.[4] Directly or indirectly, this will require the cutting down of 100,000 hectares of rainforest and, therefore, the addition of 160 million tonnes of CO_2 to the atmosphere. It is not, therefore, surprising that in 2014 the rate of deforestation jumped by 150 per cent, second only to Bolivia's jump of 162 per cent.[5]

Indonesia's record on deforestation is worse than Malaysia's or Brazil's. At the beginning of the 20th century, forests still covered 170 million hectares, or 84 per cent, of the land area of the Indonesian archipelago. During the first half of the 20th century, Dutch businesses cleared 2.5 million hectares of forests to establish plantations of sugar, rubber, teak, coffee, tea and palm oil. But deforestation accelerated in the 1970s when Indonesia decided to emulate what Malaysia's government was doing to turn it into the palm oil capital of the world. By 1990, forest cover had decreased to 128 million hectares, and the rate of deforestation continued to increase as logging licenses covering tens of millions of hect-ares and forest conversion permits whereby lands were allocated to agricultural development and settlements were routinely handed out to the crony capitalists who eventually led the Suharto govern-ment to its downfall. Forest cover therefore decreased by almost 2 million hectares a year from 128 million hectares in 1990 to 99 million hectares in 2005. The relentless burning of the felled trees during the dry winter months that follow the south-east monsoon, and the east Asian typhoon season, is one of the principal causes of the Asian brown cloud that now regularly covers South Asia and neighbouring regions from November till may every year.

Indonesia is the world's largest producer of crude palm oil and exports three quarters of what it produces, mainly to the EU where it is used both as a food and a feedstock for biodiesel production. But in recent years Indonesia has itself emerged as a major producer of biodiesel. By 2012, it had more than doubled its production from 731 million kilolitres in to 1.52 billion kilolitres (1.27 million tonnes). In 2013 and 2014, however, the slackening of international oil prices caused the production to stop growing. But following a hefty increase in production subsidy announced by the Indonesian government in February 2015, it has begun to rise again. If production reaches the original target of 3.4 million kilolitres, it will be the expense of another 400,000 hectares of rainforest cleared to plant palm oil groves.[6]

This will not end the deforestation being carried out in the pursuit of so-called 'low carbon' technologies. The FAO and the OECD have predicted a near-doubling of palm oil production by 2030. This will require the deforestation of another 4 to 6 million hectares, most of which will occur in Southeast Asia.[7]

Obtaining biodiesel, bio-methanol or gasoline from crop residues and MSW via gasification can arrest this inhuman pro-cess. Like Brazil's recent experience with ethanol, Indonesia's with palm diesel reveals the sensitivity of output to prices when a food crop is diverted to producing biofuels. The scope for doing this is ultimately as limited for palm oil as it has proved for ethanol. Fischer–Tropsch diesel will therefore supplant palm oil diesel. The only question that remains is whether this will happen in time to prevent one cause of desertification from replacing another.

Rejoining agriculture and industry

A cleaner, greener world; a narrowing income gap; new opportunities for investment and employment, and a solution to the perennial problem of agricultural subsidies, all this sounds too good to be true. But one has only to step back from the immediate present to see that these will be the consequences of healing a much deeper wound in human society that was inflicted when industry ended its dependence upon wind, water for energy and agriculture for raw materials and transferred this dependence to fossil fuels and minerals.

No previous energy shift had done this. In Europe, when the harnessing of wind and water power to production in the 10th and 11th centuries created surpluses that needed to be exchanged, Northern Europe saw the rise of the hundreds of towns of the Hansa within the space of a single century. Southern Europe saw the rise of city states in Italy and of extraordinarily wealthy cities such as Carcassonne, Beziers and Poitiers in southern France. The intra-European trade that ensued brought the fine products of the orient—spices, fine silks and cottons, gold and jewellery—to Northern and Western Europe and further raised living standards. Rapid improvements in the weaving, dyeing, printing and processing of woollen cloth in Northern Italy caused a sheep farming revolution hundreds of miles that led to the first enclosure movement in England. Similar changes were taking place in China and India. Indian cottons and Chinese silks became coveted throughout the Western world. By the middle of the 18th century, a 'putting out' system of manufacture had become established in all the great textiles weaving centres of India, and a class of traders-cum-financiers had come into being, the richest members of whom far outstripped in wealth their Venetian and Florentine counterparts.[8]

The constant feature of all these changes was the close link between agriculture and industry. This is the link that the switch from wind and water to fossil fuels broke in the 18th century. Over the next 150 years, industry's upstream links with agriculture weakened, while those with the mining industry grew ever stronger. Gradually, agriculture was reduced to the status of a service industry, necessary for human survival but not for its enrichment. The share of agriculture in the GDP fell steadily as did the proportion of people living off the land. In the cities, the gap between rich and poor—not only in terms of real incomes but also in the security of their lives—widened constantly till it forced the poor to band themselves to fight for their rights.

Replacing fossil fuels with solar and biomass energy will rejoin the umbilical cord between industry and agriculture that was severed when Europe shifted from wind and water to coal. When crop residues, and wood derived from energy forestry, become the feedstock for manufacturing transport fuels and petrochemicals, rural incomes will rise dramatically and the urban–rural gap that has become so pronounced in the low- and middle-income countries will start to close.

Since biomass cannot be transported economically for long distances, small- and medium-sized towns and cities will become the centres of energy production. These will develop into new industrial centres and lead to a renaissance of small towns. The jobs lost in the vast cities will be recreated in them, and the gap between city and countryside will eventually all but disappear. Solar energy will further shift the centres of energy production and consumption out of the cities and put them back where they were during the age of wind and water, in the heart of the countryside. In sum, by changing the 'base' of human society, harnessing the sun will change every facet of its current 'superstructure'. The benefits from the shift from fossil to solar energy, described above, will be a beginning, not an end.

A race against time

One final question remains to be answered: Can humanity dispense with fossil fuels and harness the sun in time to avert catastrophic climate change? The stock response is undisguised scepticism, for business has paid little heed to the warnings of scientists and already committed itself to making, by 2035, most of the investment for which there was space till 2100. The scepticism is not justified for, as Bruce Podobnik, the author of a recently released book titled *Global Energy Shifts: Fostering Sustainability in a Turbulent Age* has pointed out, such doubts have been aired before and have turned out to be unfounded. The example he gives is that of Britain.

> With hindsight we can see that nineteenth century analysts were wrong to claim that there were no alternatives to coal. How were they to know that a new resource—oil—would ease Britain's energy crunch? After all, it played almost no role in providing energy in that society in the 1870s and 1880s.... Although there can be no guarantees about what will happen in the coming decades, there are many reasons to believe that shifts of the magnitude and speed achieved in earlier eras can again sweep through the global energy system ... [for] ... entirely new systems of energy production, transportation, and consumption have repeatedly enveloped the world in a period of about fifty to sixty years.

History can repeat itself and do so even more rapidly than it did a century ago. Solar—particularly solar thermal—energy and biofuels can be to the world in the 21st century what oil was to the OECD countries in the 20th century. The technologies described in the previous chapters have crossed only the threshold of the possible, for in each of them we are at the beginning of the learning curve that brings costs plummeting down. The world has, therefore, already crossed a Rubicon that it did not even know existed. This is the line that divides four hundred years of growing dependence upon fossil fuels from a future in which it will need to depend only upon the inexhaustible power of the sun. The next energy shift will, therefore, be the world's last.

For decision-makers in all countries, this will open up a world of possibilities that they did not believe existed and therefore did not try to create. Far from having to cling desperately to fossil fuels for as long as possible, from having to fight for rare earths and hold international conclaves that end in mutual recrimination, heightened suspicion, muted threats and the secret hatching of contingency plans to use trade and financial sanctions against 'rogue' nations, they will have the opportunity to cooperate with each other to prove and disseminate these technologies and, in so doing, lift the threat of sudden, catastrophic climate change that hangs over the planet.

This is not wishful thinking. From the dawn of civilisation about 7,000 years ago, it took human beings 4,000 to harness animal power, as a mode of transport. This was made possible by the invention of the wheel and axle, which allowed humans to construct carts and chariots for transport, travel and war. It took another 2,000 years to advance from human and animal to wind and water power. This was the energy shift that sowed the seeds of capitalism in Europe.[9]

But the advent of markets created fierce competition. To compete, human beings had to innovate, so the speed of technological change began to pick up. So the next shift in the energy base of society, from wind and water power to coal, began only five centuries later and took only two to complete. Since this technological breakthrough occurred in Northern Europe, the resulting surge in productivity laid the foundation for Europe's domination of the world in the 18th and 19th centuries.

But by the end of the 19th century, capitalism had undergone a profound transformation. A world of small competing manufacturers, most of them little more than artisans, had given way to a world increasingly dominated by large corporations owned by thousands of anonymous investors and managed by professionals.

Joseph Schumpeter, the great sociologist and economist of the early 20th century, identified this change as the *institutionalisation of innovation*—a transfer of the search for increased efficiency from the factory floor to specialised research establishments. This freed innovation from immediate need and gave it a life of its own. The acceleration this brought about in technological change was reflected by the speed of the next and most recent energy transition, from coal to oil and natural gas. By 1960, it was not only the United States and Japan that were relying mainly upon oil, but most of Western Europe as well.

But the pace of technological change has continued to accelerate. The post-war 'golden age' of capitalism saw yet another sharp acceleration as giant corporations harnessed a spate of technologies that had been developed during the 1930s and the Second World War, but had remained unused because of the depression, or because they had been pre-empted for military use. The sheer speed at which this happened pushed Europe's post-war growth up to hitherto unprecedented rates of 5 to 6 per cent a year for more than two decades. This happened *before* the dawn of the computer age.

Computer simulation has shortened the time it takes to scale up a promising technology from laboratory to factory floor to an extent that had not been imagined, three decades ago. For instance, it used to be an engineers' rule of thumb that a new process successfully proved in a pilot plant, could be scaled up by at most four times without encountering unanticipated problems. Today, that rule of thumb is 8 and in some applications 16 times.

The meteoric rise of Japan, Taiwan and South Korea in the 1960s and 1970s, the rest of east and Southeast Asia in the 1980s and 1990s, and of China since the 1980s reveals the speed that capitalism-driven economic transformation can achieve if market conditions are favourable. It would be surprising indeed, therefore, if this acceleration in the development and dissemination of technology did not cut the time required for the next energy shift down by half.

Notes and references

1 IPCC 5th assessment Report executive summary. Table SPM-2 page 23.

2 Michael Grunwald, 'The Clean Energy Scam'. Also see Chapter 4 above—Harnessing the Sun.

3 EPEA, 'Food or Fuel: A Scientific Analysis of Environmental and Social Impacts of First Generation Biofuels' (Hamburg: EPEA Internationale Umweltforschung GmbH, May 2007).

4 Ron Kotrba, 'Malaysia Expands Biodiesel Program, Looks Beyond Five Percent', *Biodiesel Magazine* (25 July 2013), available at http://www.biodieselmagazine.com/articles/9239/malaysia-expands-bio-diesel-program-looks-beyond-5-percent-blend (accessed on 30 September 2013).

5 Carol Yong, SACCESS and JKOASM, 'Deforestation Drivers and Human Rights in Malaysia' (Forest Peoples' Programmes, 4 December 2014), available at http://www.forestpeoples.org/topics/climate-for-ests/publication/2014/deforestation-drivers-and-human-rights-ma-laysia (accessed on 14 September 2015).

6 Indonesia biofuels annual 2012 GAINS report no. ID 1222 (14 August 2012), available at http://gain.fas.usda.gov/Recent%20GAIN%20Publications/Biofuels%20Annual_Jakarta_Indonesia_8-14-2012.pdf (accessed on 11 July 2017).

7 EPEA, 'Food or Fuel'.

8 Fernand Braudel, *Capitalism and Civilisation*, vols. 1 and 3. Giovanni Arrighi, *The Long Twentieth Century*. New York: Verso (1994). For details of the wealth of Indian traders in the 16th and 18th centuries, see Prem Shankar Jha, *Crouching Dragon, Hidden Tiger: Can China and India Challenge the West?* (Counterpoint Press; New Delhi: Penguin Books [Forthcoming]).

9 Fernand Braudel, 'The Perspective of the World', in *Civilization and Capitalism (15th–18th Century)*, vol. 3 (London: Collins/Fontna Press, 1988), 93–94.

Chapter Thirteen

Catch Up—India as a Case Study

The sheer diversity of India's problems makes it an ideal showcase for the many, many ways in which a shift out of fossil fuels will transform the future of the 'South'.

The abundance of fossil fuels in the global North, and the control over the supply and marketing of oil that the industrialised countries established in the 20th century, created the base for their political and economic hegemony. The shift from fossil fuels to solar and biomass energy will partly redress the imbalance because both are far more abundant in the South.[1] How will this change the lives of the impoverished people of the South? One way to answer this question is to see what can happen in a large, low income country like India.

In the last week of October 2016, New Delhi, the capital of India, was enveloped in the worst smog that anyone could remember. What made it memorable was that it had come almost a month earlier than normal, when winter had not set in and the thermal inversions, that bring the smog generated during the day back from the upper layers of the atmosphere to the rapidly cooling earth at night, had not yet begun. Residents of the city could not even blame the festival of Diwali, when India sets off millions of tonnes of firecrackers to celebrate the victory of good over evil, because this particular smog had descended upon the city several days ahead of it.

This smog was caused by the burning of tens of millions of tonnes of stubble in the paddy fields of Haryana and Punjab where farmers were clearing their fields for planting the winter wheat crop. An annual descent, it had grown thicker every year with the size of the harvest, till it had dwarfed the effect of vehicular and industrial exhausts.

Burning crop stubble causes only severe seasonal spikes, but the two worst causes of year-round pollution are the all-pervasive habit of burning dried leaves and garbage in the open air and vehicle exhausts. Delhi is among the half dozen most polluted cities in the world. A study of air pollution and GHGs in Delhi conducted for the state government by two professors of the Indian Institute of Technology Kanpur found that particulate pollution is the main cause, and that the concentration of both PM10 and PM2.5 (coarse and fine particulate matter) is 4–7 times higher than the national air quality standard in both summer and winter months.[2]

All three main causes of Delhi's (and all Indian cities') air pollution can be eliminated by the biomass gasification technologies described earlier. The conversion of stubble into biochar will end its burning in the fields; the conversion of garbage into synthetic natural gas and transport fuels will give it an economic value that will ensure that not a single kilogramme of garbage is allowed to lie in the streets uncollected and the use of synthetic natural gas and methanol as urban transport fuels will eliminate particulate matter entirely as they yield only CO_2 and steam as their exhaust gases. Finally, gasification will eliminate dioxins and furans, the main cause of asthma and other lung diseases, and an important contributory cause of cancer.

Profoundly important though these are, they will be the lesser part of the benefits that the people will derive. The larger part will accrue to the rural population of the country. Instead of having to burn their crop stubble and incur fines for doing so, farmers in

Punjab, Haryana and every other rice-growing part of the country will be able to convert it into biochar, use a small part to rejuvenate their fields, briquette and sell the rest to companies producing bio-fuels. A detailed study of revenues from the conversion of bagasse and sugar cane waste into transport fuels presented later in this chapter shows that this will at least double the earning of farm-ers from the sale of their crop. Nationwide use of this technology will alleviate a deepening crisis in agriculture that has been driving thousands of farmers to suicide every year.

A laggard in climate change

But the benefits do not end there. As the debates during the Paris climate summit showed, India remains a late starter in the battle against climate change. As one of the two principal beneficia-ries from the exemption of developing countries from the CO_2 reduction targets of the Kyoto Protocol, it has devoted more of its energy to fighting a rearguard battle to protect its Kyoto priv-ileges than to looking for ways to reduce its emissions. The first change in its position took place at the Hokkaido summit of the G8 in 2008, when Prime Minister Manmohan Singh committed India to never exceeding the per capita emissions of the OECD countries.[3] His statement was initially misunderstood but it meant that the faster the OECD countries brought down their emis-sions, the lower would be the emission ceiling that India would be willing to accept. But seven years later it is apparent that the Indian government does not have the faintest idea of how it can live up to its promise. For, till five years ago, when SPV energy began to take on a life of its own, all it had thought of doing was blend 5 per cent of ethanol with gasoline, extract an oil akin to diesel from *Jatropha*, a plant that grows wild in India's semi-arid regions, and offer large subsidies to private companies to set up first-generation solar power plants in the country.

None of these initiatives has made the slightest dent on its dependence on fossil fuels. The blending of ethanol with gasoline has reached only 3.7 per cent and, by June 2015, India had added only 3,300 MW of solar-generating capacity to the 700 MW that existed in 2008.[4] Biodiesel from *Jatropha* has quietly slipped into oblivion.

In the meantime, its demand–supply imbalance in commercial energy has continued to worsen. While demand has continued to grow at 4 per cent a year, supply has grown at only 2.9 per cent. India has therefore become increasingly dependent upon imports which now account for nine-tenths of the oil, a fifth of the natural gas and, surprisingly, 40 per cent of the coal that the power stations consume every year.[5] These three commodities suck up almost half of its exports and have pushed the country into a chronic balance of payments deficit.

Air pollution, mounting filth and spread of disease

An immediate consequence of the growing gap between demand and supply has been a worsening power shortage. This has caused erratic voltage, constant load shedding and a growing absolute shortage of power at peak-load times. Industries have therefore set up captive power plants that work on diesel and heavy fuel oil, which now meet a quarter of their demand for power at three to four times its grid cost.

A second consequence is the drastic worsening of air pollution. A study by the World Health Organization of ambient air pollution in 1,600 cities of the world showed that Delhi had overtaken Beijing as the most polluted city in the world. What is more, 12 of the 20 most polluted cities in the world are now in India. The study ascribed this to a runaway increase in automobiles, especially those running on diesel. The study also found that air pollution had become the fifth largest killer after high blood pressure, smoke from cooking fires in the villages, tobacco smoking and malnutrition and had caused 620,000 untimely deaths in India in 2010.[6] The main causes were stroke, chronic obstructive pulmonary disease, ischemic heart disease, lower respiratory infections and cancer of the trachea, bronchi and lungs.

A third consequence is all-pervasive filth in the cities, caused by their failure to dispose of the garbage that they generate. In 2013, India had 61 cities with more than a million inhabitants. These were home to 200 million of its 360 million urban dwellers. All of these cities are choking on their own garbage. Paper and plastic bags, plastic bottles, soiled newsprint and every other conceivable sort of waste litters roadsides and railway tracks for dozens of kilometres in, and around, every town and city in the country. Most have tried to bury their garbage in landfills. But they have run out of land, so the garbage is being stacked higher and higher upon the existing landfills, until they have become stinking, smoking, man-made mountains. The waste generated by Delhi's 14 million residents now forms two hills, each 200 feet high and covering more than a hundred hectares. Roads have been carved across their slopes upon which dumpsters crawl with their loads of garbage in a grotesque reversal of what they do in opencast mines. Despite this, Delhi is failing to collect all of the garbage. In the shanty colonies, a good portion is now left to rot in the streets, be swept into rivers and open wells by the next rains or dry and crumble into dust to be blown by the wind into a million lungs. In both forms, it devastates the population with diseases ranging from gastroenteritis to asthma and cancer.

SPV and solar thermal power plants are the only kind that can increase the output of electricity rapidly enough to enable supply to catch up with demand in the foreseeable future. Biomass gasification can end India's dependence on imported oil and protect it from oil price shocks in the future. But both technologies can do much, much more.

Garbage is wealth

On 15 August 2014, in his first Independence Day speech from the ramparts of Delhi's historic Red Fort, the then newly elected

Prime Minister Narendra Modi announced his intention to create a *Swachh Bharat*—a clean and shining India. Shortly afterwards, municipal vans began to go around every area of the Indian capital playing a recorded exhortations to the residents to pick up all the garbage in their neighbourhood and deposit it in the large trash bins provided for the purpose. Other state capitals followed, but three years later the improvement has been minimal. Outside the VIP areas of the cities, garbage has continued to pile up along every road, or path, rotting in the blazing sun till it is eaten by stray cattle and pigs or turns into dust and is blown away.

What does not turn to dust is plastic, which has become the bane of peoples' lives. And not just people: one of its principal victims is India's holy cow. Banned from selling their overaged cattle to slaughterhouses, India's millions of farmers and milkmen take them into adjoining towns or wastelands and leave them to fend for themselves. Large numbers survive by foraging in garbage dumps to survive. Sooner or later they eat the plastic that has become a steadily larger portion of the garbage and die unimaginably prolonged and painful deaths.

In Bengaluru, capital of the state of Karnataka, surgeons enlisted by an NGO, Animal Aid Unlimited, operated upon an abandoned bull and extracted 20 kg of plastic from its stomach to save its life.[7] In another similar operation, surgeons of another NGO, the *karuna* (pity) society, removed 53 kg of plastic from the stomach of a cow.[8] The plastic had accumulated over 10 years in the first compartment of their stomachs because, being indigestible, it could not pass to the second. Today, there is hardly a cow or bull in the streets of the cities that has not ingested some plastic. When their stomach begins to swell it is often mistaken as a sign of health or confused with pregnancy. But the end is inevitable, and it is horrible.

In most cities, garbage removal contractors routinely dump the refuse they pick up in the central parts of the city in the adjoining countryside. This has become a threat to wildlife. In December 2016, the residents of Guwahati, capital of the north-eastern state of Assam, were shocked to learn that 26 majestic Greater Adjutant Storks, an endangered species, had died in the Deepor Beel Bird Sanctuary, adjoining the city, after eating garbage. The investigation that followed revealed that municipal garbage collectors had been dumping the garbage in the sanctuary for years and that each year had taken its toll of the stork population of the sanctuary.

No municipal authority in India has realised so far that garbage is not a nuisance but a form of wealth. This has made collection lackadaisical and disposal, an unwelcome duty to be performed at the lowest possible cost. New Delhi's fumbling efforts to deal with it reflect this paucity of imagination. The city generates more than 10,000 tonnes of garbage every day and has run out of space for landfills. Since none of the surrounding state governments is in a mood to lease it any additional land, it has turned, in desperation, to incineration. In her budget speech for 2013–2014, Delhi's former Chief Minister Sheila Dikshit proudly announced that one incineration plant had come on stream and was disposing of almost 1,950 tonnes of garbage per day.[9] Two more were being set up to incinerate another 4,300 tonnes a day and generate 50 MW of power.

Other cities had been watching Delhi's initiative, so when, a few days later, the federal (federal) government unveiled substantial subsidies for 'waste-to-energy' projects in the budget for 2013–2014, there was a deluge of proposals from private companies. Within months, several state governments had applied for, and obtained, the approval of the Central Ministry of New and Non-Renewable Energy for eight more waste-to-energy projects. All of these were given for incineration plants. When the city of Chennai, in Tamil Nadu, received a lone proposal for waste gasification (along with six others for incineration plants), the local authorities dismissed it out of hand as being too expensive.

Neither the national nor the state governments seem to be aware that they are being deluged with proposals to set up incineration plants just when cities in the high income industrialised countries have begun to phase out an entire generation of them because of the deadly threat they pose to human health. The threat comes not from carbon particles, or nitrogen and sulphur in the flue gas, but from another category of pollutants known as dioxins and furans. Dioxins are stealthy killers. The word is a generic term for more than a hundred very long-lasting chemicals that are produced by burning municipal and medical waste. Their distinctive feature is that they cannot dissolve in water, so they settle on land and water bodies and are absorbed in their entirety by terrestrial and aquatic vegetation. From this, they travel up the food chain into the animals and fish that feed on the plants and thence into humans. Since living organisms cannot break them down they are found in very high concentrations in meat, fish, milk and eggs. In human beings, a prolonged exposure to dioxins—through what we call a 'rich diet'—impairs the functioning of the liver and the immune and reproductive systems and raises the incidence of cancer.

In sum, in ways that we are only now beginning to understand, dioxins shorten our life span. The US Environmental Protection Agency, which put together the first comprehensive report on dioxins in 1994, described them as 'the most poisonous substances known to man'. Not surprisingly, a study by the Paris-based Institut national de la santé et de la recherche médicale, released in 2000, revealed that waste incineration plans contributed 46 per cent of the dioxins in the atmosphere in France and 87 per cent in Germany.[10] Men have absolutely no way of expelling them. Women can do so, but only by passing them to foetuses in their wombs or breastfeeding their babies.

Dioxins get broken down into their basic, non-toxic, elements at temperatures above 850°C, but maintaining this temperature all the time is made difficult by the variability of the composition of untreated solid waste and of its moisture content. Even when temperatures above 850°C are maintained, the need to cool the flue gas before it can be cleaned allows some of the dioxins to get re-synthesised using dust particles as a catalyst.

Modern incinerators have found a way around the second problem but not the first. In Finland, the government ordered the temporary closing down of an incineration plant built with the most elaborate safeguards when it found, after two years of its operation, that dioxin levels in the surrounding vegetation had risen by 15 to 25 per cent within a distance of 4 km from the plant. This problem is especially difficult to resolve in developing countries because of the high proportion of kitchen waste and the abrupt day-to-day, sometimes hour-to-hour, changes in moisture content caused by rainfall.

Throughout 2012 and 2013, whenever environmentalists pointed out the acute hazards of incinerating garbage, the Delhi government assured them that elaborate safeguards had been incorporated into the design of the three plants under construction to ensure that they met the prescribed environmental safety norms. One of these was the replacement of raw waste with pelletised RDF extracted from it to guarantee an even quality of fuel for the incinerators. But test after test showed that this had not solved the problem. In tests carried out in March 2012 at Delhi's Okhla plant, the total particulate emissions from its two chimney stacks stayed within norms on six occasions but exceeded them on four. However, the emission of dioxins was 30 times the maximum permissible limit. The company promised to rectify the defects but in a second test in May 2013, designed specifically to check the release of dioxins, the flue gas from its two stacks was found to contain 2.8 and 12.7 times the permitted maximum of dioxins and furans! But despite this, the plant was not immediately closed down because the alternative was to find landfill space for 2,000 more tonnes of waste every day.[11] The plant was closed down in 2016 when a new government came in to power in Delhi state, but the National Green Tribunal supported its reopening before the court on the grounds that it had again passed the test for particulate matter in the flue gas. Dioxins did not even get mentioned.

The obstinacy of the former Delhi government had provoked accusations of corruption and crony capitalism. But the underlying cause was the mindset that garbage is a nuisance and has no value, so contracts for waste-to-energy projects must go to the lowest bidders. This meant that tenders were appraised solely on the basis of their capital and operational costs and not their revenue-earning potential. As a result, the Okhla plant was built for a mere $44.6 million, one-tenth of the cost of equivalent-sized plants in Europe.[12] On this basis, gasification has not stood a chance.

No one in the outgoing administration seemed to know that India already had two small plasma gasification (68 tonnes-a-day) plants set up to destroy hazardous wastes at Pune and Nagpur—two large cities in the industrially advanced state of Maharashtra. These were among the first plants set up by Westinghouse anywhere in the world and had been supplying process heat and raw fuel gas to a neighbouring power station for several years.

Self-sufficiency in transport fuels

India now stands at a crossroads. By 2025, over 400 million Indians will be living in cities with more than a million inhabitants and will be generating 400,000 tonnes of garbage a day. It can continue blindly down the road to incineration. But that will seal the fate of its city dwellers, especially children, whose lives will be cut short by diseases that can easily be avoided. Alternatively, Indian cities can sort their garbage to remove the inorganic portion of the waste and obtain up to 100,000 tonnes of dry refuse-derived fuel from the remainder and convert it into 35,000–50,000 tonnes of transport fuels, depending on the technology they choose, *every day*.[13]

This would reduce India's oil imports by 10 million tonnes a year, but that would only be a beginning. For if India's sugar industry decides to co-produce a 'Fischer–Tropsch' transport fuel in addition to the ethanol it is already producing, it will produce 10 times this amount of transport fuels every day. That will make India almost self-sufficient in transport fuels and give the sugar industry a new lease on life at a time when it is finding it more and more difficult to make both ends meet.

India's sugar industry is second only to Brazil's in size. But for the better part of two decades, it has enjoyed a precarious existence because its revenues have been squeezed relentlessly by price controls on a large part of the sugar it produces and annual increases in the minimum price it must pay to farmers for their sugar cane. The industry has stayed afloat so far by increasing its productivity and output of by-products, but this has become more difficult year by year. The first step it took, at the end of the last century, was to replace outmoded conventional boilers that ran on coal with modern high pressure boilers that ran on bagasse, the residue left after squeezing the juice out of sugar cane. This not only saved sugar mills the expense of purchasing and transporting coal, but yielded four times the heat they needed to distil the cane juice and crystallise the sugar. They used the surplus heat to generate power and sell it to the state electricity grid.

Starting from one mill in 1993, 211 of the country's larger sugar mills had installed modern boilers by 2010 and were feeding the bagasse obtained from 240 million tonnes of cane into them. This enabled them to install 3,321 MW of power-generating capacity and sell the electricity to the state grids for 180 to 200 days a year. This should have eased the industry's plight but did so only partially because, as the global recession deepened and India's growth rate dropped, state governments began delaying the payment of their electricity dues, often for years at a time.[14]

To ease the crisis, the spokesmen of the Indian Sugar Manufacturers' Association urged the central government to lift, or at least ease, price controls on 'levy' sugar and raise the 'mandated price' that oil marketing companies had to pay for their principal by-product, ethanol. As of mid-2015, the industry had not considered replacing its older boilers with gasifiers and producing diesel and ATF in addition to ethanol. But doing so would not only boost the industry's revenues four or five fold, but turn it, almost overnight, into the most important industry in the country.

This is because they would be able to gasify not only the bagasse they now burn, but also other sugar cane waste (leaves, roots and residues from the production of ethanol such as press mud).[15] Every tonne of sugar cane that is crushed yields about a third of a tonne of bagasse. But every tonne of sugar cane also leaves behind a tonne of leaf and roots. Thus, for every tonne of bagasse, which the sugar mills are burning in their boilers today, they can gasify four tonnes of pelletised feedstock. This will make it possible to operate the plant for the whole of the year instead of only for six months and allow the managers to double the capacity of the gasifiers they install.

Some idea of how much the industry stands to gain by producing a 'Fischer–Tropsch' transport fuel in addition to sugar and ethanol can be had by comparing the sales revenue of a small, conventional mill that crushes 4,000 tonnes of cane a day, burns bagasse in its boilers, produces electricity and sells ethanol and power as by-products, with a similar mill that gasifies a pelletised mixture of bagasse and sugar cane waste, and produces synthetic transport fuels in addition to ethanol.

At 2014 prices of sugar, ethanol, methanol, gasoline, diesel and electricity in India, the sales revenue of the first plant would amount to $63 million a year.[16] But the second would generate more than $300 million, because it would be producing not only as much sugar and ethanol as before but also close to 400,000 tonnes of transport fuels every year.[17]

The benefits for India's economy would be staggering. In 2014–2015, transport fuels and LPG (which can be replaced by DME) accounted for 113 out of the 165 million tonnes of oil consumed in the country. Allowing for refinery losses, this amounts to 70 per cent of consumption and 87 per cent of imports in the same year. If this is replaced by biofuels, India can save over $100 billion in import bills, and its balance of payments will begin to show a healthy surplus.

The rise in the value of the sugar cane crop will at least double sugar cane prices. The sale of their cane waste will add another substantial amount to their revenues.[18] More than a hundred million families would face a transformed future. This one industry, therefore, has the capacity to make India largely self-sufficient in transport fuels, reduce its import bill by a quarter, shield it from future oil price shocks and transform the future of agriculture.

What solar thermal power can mean for India

Solar power has so far been considered capable of meeting demand only during daylight hours. But the breakthrough in solar thermal power generation described earlier makes it capable of replacing not only coal but also hydro and nuclear power. India derives 56 per cent of its commercial energy from coal. Its current consumption is over a billion tonnes a year and is expected to double by 2020. Between November 2014 and May 2015, the central government cleared 41 new coal-mining projects. If these develop as planned, India will be adding one new mine every month to six weeks from 2016 till 2020.[19] If avoiding trade sanctions were to remain the only reason for reducing coal's preponderance, it will negotiate hard and delay doing so for as long as possible. But lowering CO_2 emissions should be the least of its reasons for finding alternatives to coal. No country has as much to gain from harnessing solar thermal power as India does. India has been suffering from an endemic power shortage, peaking at 13.5 per cent of its total requirement in some seasons of the year and has been unable to bridge the gap despite hectic efforts for the past four decades. The main reason is that its hydroelectric potential is locked up in remote regions of the Himalayas and despite having the fifth largest coal reserves in the world it is unable to get to them because most of the coal is located under land that is home to the poorest and most neglected people in the country. These have been fighting expropriation in every way possible for the last half-century, first in the courts and now with guns. As a consequence, although the central government issued almost 200 mining concessions to private firms between 1993 and 2011, almost none of the concessionaires have been able to mine any coal in these lands.

In 2011, therefore, private power plants with 35,000 MW of generating capacity were standing idle for want of an assured supply of coal. To remedy this, the Indian government has been importing coal from Indonesia, South Africa and Australia. But the shortage has now become endemic. As a result, the average PLF (also called capacity factor) in Indian coal-based power plants fell from 79.7 per cent in April to June 2009 to 68.7 per cent in the same quarter of 2013[20] and has fallen further in 2015 to 65 per cent.[21]

Concentrated solar thermal power can solve all of these problems. As shown by the Ouarzazate plant in Morocco, power plants that supply power throughout the day with very limited fossil fuel backup can be up and running in a little over two years, against the six or seven that coal-based plants are taking from the date of conception.[22] In sharp contrast to coal-based plants that are now fed almost entirely by open cast mines that tear open huge swathes of land, solar power plants do not need mines and only minimally disturb even the land they are built upon. In India, solar power projects are experiencing none of the problems in finding land that the coal mines face, because the land that suits them best is often the land least fit for cultivation or animal husbandry. In the United States, nearly all the solar power plants being built today are in the deserts of California, Arizona and Nevada. In India, nearly all are being planned for the near-desert regions of Gujarat, the Thar Desert in Rajasthan, semi-arid parts of the Deccan Plateau in southern India and the high mountain deserts of Ladakh in Kashmir. Not only are there no forests to be felled, rivers to be protected, villages to be moved en bloc or deep trenches to be cut for open cast mining, but much of the land has no other use. A shift to solar power will make it unnecessary to deprive the forest-dwelling tribals and other landless poor in central India, who have customary rights but no legal title to the land on which they live.

Among private investors, interest in solar thermal power is rising rapidly because of the steep fall in capital cost after the entry of the Chinese into the market for solar panels and heliostats. But its main contribution will be to society, for it meets every one of the six conditions that have been set out in Chapter 3, for gaining acceptability as a replacement for fossil fuels: it is a proven technology on the verge of becoming cost competitive; will cause no social disturbance and will fit into the existing energy infrastructure like a hand in a glove. It is also the only source of power whose supply can outpace the rise of demand and thus end India's endemic power shortage.

Unlike transport fuels, this message has not been lost upon the Indian government. In the second phase of its National Solar Mission, it had set a target of 22,000 MW for 2022, but in 2014 the newly elected Prime Minister Narendra Modi raised the target to 100,000 MW. But Indian investors have been eyeing SPV power.[23] In May 2015, businessmen accompanying him on a visit to China signed a number of agreements with Chinese companies, but all of them were for SPV power. The construction of the world's largest SPV plant, with a capacity of 4,000 MW, was completed in Rajasthan in 2016, and two more similar giant plans are under construction in other states. Several enterprises have explored solar thermal power. In November 2013, the Gujarat State Electricity Corporation Limited invited expressions of interest from consultants for designing a 50 MW solar thermal power plant with sufficient heat storage to provide power on demand, but very high real rates of interest have both increased the capital cost of solar thermal power plants and increased the risks to the entrepreneur if construction gets delayed. As a result, little has been heard of the project since then.[24] As of late 2015, there was one faltering, early generation plant in southern India with no provision for heat storage and no commercial plants under construction.

How even the most endangered countries will benefit?

Other developing countries will reap similar benefits in differing proportions depending upon their climate and topography. The prime beneficiaries will be the countries of the swathe of most endangered countries that spreads across south-western Africa, the Sahel, the African littoral of the Mediterranean and the Middle East. As Morocco has already shown, solar thermal and PV power will not only reduce the dependence of arid and desertified countries on imports to meet their energy needs, but be able to export electricity to the countries of the North to help them reduce their dependence upon coal and natural gas. But, in human terms, using MSW and crop residues to produce transport fuels will do far more. For whereas drought and freak weather events often destroy cereal crops, unless they are exceptionally severe they leave the straw, stalks and stubble largely unaffected. This gives the farmer one valuable crop he or she can still sell for a good price. Desertification will therefore largely cease to force mass migration to towns or other less affected areas of the country or the world.

Averting calamity in the Himalayas

Solar energy's greatest contribution would be to replace hydroelectric power projects altogether. Environmentalists, seismologists and human rights activists have been questioning the utility of hydroelectric power projects for several decades but so far the engineering and heavy industry lobbies have always won the day. Indian planners routinely claim that hydro-power is an environment-friendly alternative to thermal power; that it is needed to meet peak loads and thereby maximise the efficiency of thermal power plants, and that although their capital cost is high, their operating cost, when compared with coal-fired plants, is very low.[25] China has justified the renewed emphasis on hydro-power projects in its 12th Energy Plan on much the same grounds.

But these claims are propaganda, not fact. In 2008, the Indian CEA released a 10-year hydro-power development plan that identified a 'shelf' of 146 large- and medium-sized hydel power projects that would add 58,573 MW of generating capacity during the 11th and 12th Five Year Plans, between 2007–2008 and 2016–2017.[26] But only a year earlier a standing committee of the Indian Parliament had found that the hydel schemes implemented in the previous 60 years had costed 3.88 to 25.16 times more than the original project estimates and had taken 9 to 16 years longer to complete than anticipated.[27] Learning from this experience, the CEA estimated that future projects would be completed in approximately 10 years, of which the actual construction would require five years.[28] This has turned out to be wishful thinking for only 49 projects, with a cumulative capacity of 15,006 MW taken up by 31 March 2014. Of these, the CEA expects only 35, with a generating capacity of 9,934 MW to be completed by 2017.[29]

The popular adage that hydro-power is cheaper in the long run than coal-based power is also a myth. It has been created by using a 'standard' capacity factor of 60 per cent, or 5,256 hours a year, to calculate the cost per unit of the delivered power. But no Indian, or for that matter Chinese, hydro-power plant has ever achieved this level of capacity utilisation. In 2009–2010, Indian hydel plants generated power for only 2,893 hours in the year, giving a PLF of 32 per cent. This effectively doubles the cost of delivered power from a hydel power generation. The capacity factor of even the Three Gorges Dam power project is rated at 46.2 per cent.[30]

Finally, hydro-power can no longer be defended even on the ground that it is cheap once the dams have been built. As was shown in Chapter 6, with the cost of heliostats decreasing rapidly, the delivered cost of solar thermal power is beginning to compete with coal-based power. In addition, solar power plants take two years or a little more to construct, against 15 years for the 1,500 MW Nathpa Jhakri power plant in the central Indian Himalayas and 17 years for the 21,000 MW Three Gorges project.[31] This makes the true cost of solar power, after taking into account the additional production and savings made possible by providing the needed energy a decade earlier, saving of time, a fraction of the cost of hydroelectricity: it could even make it negative.

Even this benefit would be trivial when compared to the immense ecological, social and human damage that will be avoided by not building hydroelectric projects. In the Himalayas, this is magnified by the fact that these mountains are built upon a tectonic fault line in one of the most unstable seismic zones in the world. This is in the Himalayas, where the Indian tectonic plate crashed into the Eurasian Plate 51 million years ago and continues to grind a little more into it every year.

The most unstable part of this unstable region is the northeastern corner of the Himalayas where they and two other mountain ranges come together. It is precisely there that India and China are planning to build no less than 97,000 MW of hydro-power generating capacity. That is four and a half times as much as in the Three Gorges Dam project.

India, for its part, plans to generate 22,000 MW from two large dams on the Brahmaputra in Arunachal Pradesh and 10,000 MW from smaller dams on its tributaries. But its plans are dwarfed by those of China which intends to build 40 dams on the Yarlung Tsangpo, known in India and Bangladesh as the Brahmaputra, and its tributaries. Twenty of these dams, on the Yarlung Tsangpo itself, are expected to generate 60,000 MW of power. Nine of these are monster projects which will generate 40,000 MW of power at the 'Big Bend', where the river makes a huge 'U' turn and falls from 3,500 m on the Tibetan Plateau to 700 m, in the undulating hills of Arunachal Pradesh in India.

The plan for these, put forward by the China Yangtze Power company in which the family of former Premier Li Peng has a controlling interest, and the state-owned Three Gorges Dam Company is to build a vast tunnel under the ridge that separates the two arms of the Big Bend and divert 50 billion cubic m of water a year to the south-eastern slope where it will fall over nine cascading hydro-power dams to generate 40,000 MW of peak power.

Neither China nor India have made even a rudimentary assessment of the impact that tearing down billions of cubic metres of rock and earth to build dams, tunnels and roads, and storing millions, in some cases billions, of cubic metres of water will have upon the stability of the earth's crust in this region. This neglect seems deliberate, for both governments are fully aware that this corner of the Himalayas has seen a succession of the most severe earthquakes in recorded history. The most recent occurred in April 2015 when an earthquake that registered 7.9 on the Richter scale hit Nepal and adjoining areas of India and cost more than 8,000 lives in Nepal. But prior to this, there have been four of these mammoth quakes, measuring 7.8 to 8.7 on the Richter scale within a span of 53 years between 1897 and 1950. The first and last occurred just 53 years apart in the region immediately south and west of the Big Bend in the Brahmaputra.

The 1897 earthquake measured 7.8 on the Richter scale (equivalent to the explosion of 7.6 million tonnes of dynamite, or a medium-sized hydrogen bomb) and caused widespread damage and loss of life in what was then called upper Assam. It was caused by the build-up of pressure as the Indian (tectonic) plate pressed against the Shillong Plate, a part of the far older Eurasian Plate. The quake occurred when the former shifted 11 to 16 m as it dived under the latter, over a stretch of 110 km, along what is known as the Oldham fault. This is one of the largest tectonic shifts recorded so far. Its effects were felt through the entire earth's crust over this length.

The 1950 earthquake was the severest recorded in the Himalayas. It occurred at Rima, Tibet, not far from the site of the 1897 quake. Measuring 8.7 on the Richter scale, it is 1 of the 10 most severe earthquakes in recorded history. Its epicentre also lays on the fault line where the Indian continental plate hits and dives beneath the Eurasian Plate. Survivors from the region reported mudslides damming rivers and causing them to rise high when these broke, bringing down sand, mud, trees and all kinds of debris. Pilots flying over the area reported great changes in topography, caused by enormous landslides, some of which were photographed. An aftershock of this earthquake, a long way to the west of it, was severe enough to register 8.6 on the Richter scale and caused avalanches and floods that destroyed swathes of forest in the Mishmi and Abor Hills.

Most of the damage these earthquakes cause arise from the landslides they cause. These block rivers, causing them to rise till the pressure of the stored water breaks through. The result is a flash flood downstream that causes havoc among the villages and towns that lie in its path. The 1950 earthquake and the 8.6 magnitude aftershock that followed caused avalanches that blocked several of the tributaries of the Brahmaputra. One such dyke in the Dibang Valley broke quickly and caused relatively little damage. But another, at Subansiri, broke after water had collected behind it for 8 days and unleashed a 7 m high wave that submerged several villages and killed 532 people in what was then a very sparsely populated area. In all, the 1950 earthquake killed more than 1,500 people in Assam.

Geological studies, including the radio carbon dating of sand found on the surface, have uncovered at least one other giant earthquake in the same area that took place in 1548 and two earthquakes in the central region of the Himalayas that were severe enough to break the earth's crust all the way to the surface. These occurred in 1255 and 1934.

In terms of human lives, the 1934 earthquake, which measured 8.1 on the Richter scale and had its epicentre about 10 km south of Mount Everest, very close to the epicentre of the 2015 earthquake, devastated north Bihar and Nepal, and killed at least 30,000 people. This occurred before any dams had been built. The dykes that were overwhelmed were of mud and broke in a matter of days. But earthquakes of this magnitude will almost certainly break concrete dams as well. For the Richter scale is a logarithmic scale. An 8.1 magnitude earthquake releases three times as much energy and an 8.7 magnitude releases 23 times as much as a 7.8 magnitude quake.[32] Should any of the proposed dams crack during an earthquake or an ensuing flood, the colossal wave of water, mud and boulders that will be released will kill millions and completely devastate the areas of Tibet, India and Bangladesh that lie in its path. The overwhelming majority of deaths and damage will take place in India and Bangladesh.

India got a foretaste of this some years ago when a flash flood in the Yarlung Tsangpo, caused by torrential rains and landslides, wiped out an entire island in the Brahmaputra, killing nearly all who lived on it. Chinese hydrologists knew that the flood would occur but did not warn their Indian counterparts. India got another foretaste in June 2013 when three days of torrential rains in the valley of the Bhagirathi River, one of the two main tributaries of the Ganges River, caused landslides, which blocked a tributary and destroyed the entire town of Kedarnath when the mud dyke broke, and killed between 5,000 and 10,000 people in a matter of hours. Over the previous 20 years, the hill slopes overlooking the Bhagirathi valley had been ravaged by the construction of dams and tunnels for the Tehri hydroelectric project, the second largest in India. The rains therefore brought on a catastrophe that many had feared, but hoped would never happen.

The Tehri hydro-power project has a generating capacity of only 1,000 MW, one-fortieth of what the nine dams at the Big Bend will have and almost one-hundredth of what India and China intend to create in the region. Twenty years from now, when the thousands of explosions that have been used to build tunnels for 240 or more power projects have loosened billions of tonnes of earth and the population has increased tenfold, the entire region will become a calamity waiting to happen.

Notes and references

[1]   For instance, solar power stations get direct sunlight for an average of 22 per cent of the time in India, 1,900 hours, against only 6 per cent of the time in Great Britain. See David Rose, 'Why India Has Been Captured by Coal', *The Guardian* (27 May 2015), available at http://www.theguardian.com/news/2015/may/27/why-india-is-captured-by-carbon (accessed on 11 July 2017).

[2]   Mukesh Sharma and Onkar Dikshit, 'Executive Summary' (IIT Kanpur), available at http://delhi.gov.in/DoIT/Environment/PDFs/Final_Report.pdf (accessed on 11 July 2017).

[3]   India is among the permanent invitees to G8 meetings.

[4]   Government of India Press Information Bureau, 'Progress Under the Jawaharlal Nehru National Solar Energy Mission' (11 May 2012), available at http://pib.nic.in/newsite/erelease.aspx?relid=83632 (accessed on 15 October 2015).

[5]   Rose: op. cit.

[6]   India Environment Portal, 'Ambient (Outdoor) Air Pollution in Cities Database 2014 (WHO Study on Urban Air Quality), available at http://www.indiaenvironmentportal.org.in/content/391768/ambient-outdoor-air-pollution-in-cities-database-2014-who-study-on-urban-air-quality/ (accessed on 11 July 2017).

[7]   See https://youtu.be/YUR5T-Kx3hE (accessed on 11 July 2017).

[8]   See https://www.youtube.com/watch?v=YUR5T-Kx3hE&t=202s (accessed on 11 July 2017).

[9]   India has a federal structure with considerable powers left in the hands of the state governments. Land and urbanisation policies fall entirely within the latter's purview, but the central government can encourage them to adopt specific initiatives by allocating tied grants from its coffers.

[10] INSERM, 'Collective Experts' Reports (2000), available at http://www.ncbi.nlm.nih.gov/books/NBK7127/ (accessed on 11 July 2017).

[11] It was closed down by the government of Arvind Kejriwal, who led a new political party the Aam Aadmi Party (common man's party), to win 67 out of 70 seats in the Delhi state legislature. Kejriwal closed the plant in February 2015.

[12] David Ferris, 'Out of India's Trash Heaps: A Controversy on Incineration', *Yale Environment 360*, available at http://e360.yale.

edu/features/out_of_indias_trash_heaps_a_controversy_on_inciner-
ation (accessed on 26 March 2017).

[13] By the Solena plasma gasification process, 100,000 tonnes of RDF will yield approximately 40,000 tonnes of transport fuels per day. In other words, if all the garbage in India's million-plus cities in 2025 were converted into transport fuels it would replace 10 to 11 million tonnes of gasoline. This is half the present consumption and a third of the expected consumption in 2025.

[14] In the beginning of 2013, for instance, just two state governments in India, those of Haryana and Tamil Nadu, owed their sugar mills $30 million (Data provided by Indian Sugar Mills Association [ISMA]).

[15] Press mud is a sludge left over from the fermentation of molasses to obtain ethanol and 68 per cent of it is carbon.

[16] This estimate assumes an ex-factory price of sugar as ₹35,000 per tonne, power sale price of ₹5 per KWh, ethanol sale price of ₹35 per litre and ₹62 to the dollar. It is assumed that the plant uses its electricity to run not only the fuel synthesis plant but also an electrolysis plant to feed both oxygen and hydrogen to the gasifier. This will yield half a tonne of methanol or an equivalent transport fuel per tonne of biomass.

[17] BA's GreenSky project outside London expects to produce 200,000 tonnes of transport fuels from 575,000 tonnes of RDF obtained from London's solid waste. The same ratio has been used here.

[18] S.R.S. Murthy, 'The Economics of Sugar Cane Production and Processing' (Mumbai: National Bank for Agriculture and Rural Development, 2010), Chapter 5, 41–60, Table 5.2.

[19] Rose: op. cit.

[20] Sarita Singh, 'Power plants' Capacity Utilisation at Record Low' (New Delhi: *The Economic Times*, 8 July 2013), available at http://articles.economictimes.indiatimes.com/2013-07-08/news/40443335_1_gas-plants-plf-power-generation (accessed on 15 September 2015).

[21] Vikas Dhhot, 'India Sees Lowest Plant Load Factor in 15 Years; Power Capacities Operating at 65%', *The Economic Times* (29 May 2015), available at http://economictimes.indiatimes.com/industry/energy/power/india-sees-lowest-plant-load-factor-in-15-years-power-capacities-operating-at-65/articleshow/47463610.cms (accessed 15 October 2015).

22 Very few coal-based plants are actually being completed in this time period. The main reason is the lack of environmental clearances. This is a direct consequence of the Maoist uprising in the minerals belt of central India.

23 *CleanTechnica*, 'India Ups Solar PV Target by 30 Percent to 1,000 MW', available at http://cleantechnica.com/2014/04/07/india-ups-2014-solar-pv-target-30-1000-mw/ (accessed on 11 July 2017).

24 Project Monitor, 'GSECL Exploring Possibility of Setting Up 50 MW Solar Thermal Power Plant' (30 November 2013), available at http://www.projectsmonitor.com/daily-wire/gsecl-exploring-possibility-of-setting-up-50-mw-solar-thermal-power-plant/ (accessed on 15 October 2015).

25 Dun and Bradstreet, 'India's Energy Sector: Executive Overview'.

26 Central Electricity Authority, Hydro Planning and Investigation Division, 'Hydro Development Plan for 12th Five Year Plan (2012–2017)' (New Delhi: 2008, September).

27 Utpal Bhaskar, '16-year Delays, 2,532% Cost Overruns Shock House Panel'.

28 Central Electricity Authority, Hydro Planning and Investigation Division, 'Hydro Development Plan for 12th Five Year Plan (2012–2017)', iv and v.

29 South Asia Network on Dams, Rivers and people (SANDRP), 'Massive Hydropower Capacity Being Developed by India: Himalayas Cannot Take This Onslaught' (6 May 2014), available at https://sandrp.wordpress.com/2014/05/06/massive-hydropower-capacity-being-developed-by-india-himalayas-cannot-take-this-onslought/ (accessed on 15 October 2015).

30 CleanEnergy action Project, 'Three Gorges Dam Hydroelectric Project', available at http://www.cleanenergyactionproject.com/CleanEnergyActionProject/Hydropower_Case_Studies_files/Three%20Gorges%20Dam%20Hydroelectric%20Power%20Plant.pdf (accessed on 11 July 2017).

31 Ibid.

32 See http://en.wikipedia.org/wiki/Richter_magnitude_scale#cite_note-18 (accessed on 11 July 2017).

Chapter Fourteen

Retreat from War

Ensuring an uninterrupted supply of oil has been one of the most important causes of militarisation and conflict in the past 100 years. Today a second, even more important non-renewable resource, water, is also becoming scarce. The shift from oil and gas to biomass will eliminate the first cause of conflict. Replacing hydroelectric with solar power will greatly reduce, although not eliminate, conflict over the second.

The race to harness river waters described in the last chapter does not pose a threat only to the ecology but also the peace and security of nations. Disagreements between upper and lower riparian states over their rights to river waters have sparked many wars in the past. Today, with the human population touching 7.5 billion, and per capita water consumption rising rapidly in developing countries, the possibility of such conflicts is greater than ever. There are around 200 shared river basins in 148 countries; 37 of these countries, that house more than half of the world's population, are involved in conflicts that have ended, or could end, in war.[1] The overwhelming majority are in developing countries where the population is rising, water use in agriculture is exploding and the underground water table is falling. These countries are also starved of power. The potential for conflict is therefore even greater here than in the case of oil and other mineral resources.

While solar energy cannot augment the supply of water, it can eliminate the main cause of conflict by making hydroelectric power projects redundant. SPV plants can do this by providing peak power far more cheaply, rapidly and in an eco-friendly way, than either storage or run-of-the-river power plants. Solar thermal (CSP) plants can go a step further and take care of most of an areas need for base-load power as well.

Border dispute in the Himalayas

The way in which competition over the waters of the Brahmaputra basin has become politicised illustrates how easily this can happen. The first indication that China might be considering building dams on the Yarlung Tsangpo to divert some of its waters to the arid regions of the north came at the first international conference of the Global Infrastructure Fund in Anchorage, Alaska, in 1986. Although Chinese officials rubbished the idea as being impossibly expensive to implement, they did not rule out the possibility of constructing dams on the river to generate power. This ambivalence raised understandable alarm in Bangladesh and India but Beijing sought to allay their fears by assuring them that it intends to build run-of-the-river dams that would redirect, but not stop, the flow of its waters into India and Bangladesh.[2]

These reassurances did not, however, prevent China and India from entering an undeclared race to capture the hydroelectric potential of the Brahmaputra river basin. Chinese publications began to air plans for harnessing the Yarlung Tsangpo to generate power in 2005 but India's fears had already been revived by the publication of a book by Li Ling in 2003 titled *Tibet's Waters Will Save China*.[3] As the downstream riparian, India hoped to establish first user rights on the water in order to bolster its claim to an uninterrupted flow of the Brahmaputra's waters. In 2007, India announced that it would take up 146 projects up in the Brahmaputra basin in the next 10 years.[4]

Such ambitious but conflicting plans were bound to have a political fallout. Its first indication was an abrupt hardening of China's stand in the six decades old dispute over the border between India and Tibet in the Himalayas. This was signalled in 2006 by the Chinese ambassador to India, days before President Hu Jintao was scheduled to arrive on a state visit. In a televised interview, he stated that China considered the whole of the north-eastern Indian state of Arunachal Pradesh to be a part of Tibet. This reversed Beijing's earlier position, developed in a succession of bilateral negotiations since 1994, that China was prepared to settle for a substantial modification of some parts of the existing temporary boundary, called the Line of Actual Control. The announcement took the Indian government by surprise and was followed by three years of rising tension along the border. China began to refer to Arunachal as 'South Tibet' and to its principal monastery at Tawang as Tibet's second most important monastery after Lhasa. It also began to deny visas to Indian officials who were serving in Arunachal Pradesh. The tension was not defused till there was a meeting between Premier Wen Jiabao and Prime Minister Manmohan Singh designed specifically to prevent its spilling over into military conflict at Hua Hin Thailand in October 2009.[5] Since then, China and India have completed building major dams on the river, and have begun work on others, though not on the 'Big Bend'. This has not raised new tensions between the two governments, but if alternatives to hydro-power are not found soon the respite this meeting gave the two countries is likely to prove temporary.

SPV and thermal power can not only avert this conflict but make the provision of peak-load power into an area for economic cooperation. Chinese companies have built up huge excess capacity in the manufacture of SPV panels heliostats. It also has a large trade surplus with India that it wishes to balance with investment flows in the opposite direction. A commitment to building solar instead of hydro-power plants in the Brahmaputra basin will give both countries the power they need far more quickly and cheaply than the dams they are intending to build on the river and its tributaries.

Nor does solar energy threaten the ecology or the civilisations of the Brahmaputra basin. The Gemasolar plant in Spain requires 1 km² of land on an average for every 10 MW of generating capacity. Thus, the entire 97,000 MW of generating capacity that China and India are racing to install on the Brahmaputra and its tributaries can be set up on 9,700 km² of land. This is a quarter of the area in the Thar Desert that India has set aside for a solar power park. It is also less than two-thirds of 1 per cent of the Gobi Desert in China.

Conflicts over oil will spill to rare metals

Conflicts over water are not the only threats to peace that harnessing solar power can mitigate. A much older is the conflict over minerals. Political economists of the 19th century were fully aware that this could not go on forever and warned against letting it go out of hand. In his *The Principles of Political Economy*, John Stuart Mill wrote:

> It must always have been seen, more or less distinctly, by political economists, that the increase of wealth is not boundless: that at the end of what they term the progressive state lies the stationary state, that all progress in wealth is but a postponement of this, and that each step in advance is an approach to it.[6]

Karl Marx's 'classless society', in which the need to accumulate capital, and therefore to exploit, will have ended is also the stationary state by another name. What no one was able to chart is the way from the 'progressive' to the stationary state. So the question faded into the background till the oil price shocks of the 1970s, and the predictions of the Club of Rome, reminded humanity that it is living on a finite planet.

But the machine that drives capital accumulation cannot be stopped. So relentless growth continues to fuel the demand for energy and, therefore, a growing competition between nations do discover, and ensure access to, new deposits of fossil fuels. This growing competition over scarce natural resources underlies the 'clash of civilisations' that has replaced the clash of ideologies after the end of the Cold War. For the end of the Cold War coincided with the emergence of China as a resource guzzling industrial colossus. Its impact was hidden for almost a decade by the sharp drop in the demand for raw materials that followed the economic collapse of Russia and the east European economies. But in 2002, China's imports skyrocketed as it entered a new investment boom. With that the Indian summer of cheap oil finally ended.

Long before this, however, the growing dependence of the industrialised countries upon imported energy, notably oil, had made ensuring its uninterrupted supply an important goal of foreign policy. The United States, for instance, knew the importance of uninterrupted oil supplies for Japan's military expansion in Southeast Asia. So when Japan invaded French Indo-China in 1940, it belatedly declared an embargo on oil exports to Japan in July 1941 and cut it off from 93 per cent of its supplies. Japan responded by sending a carrier fleet to annex Indonesia, then the fourth largest oil producer in the world.[7] This led to Pearl Harbour and the entry of the United States into the Second World War.

The potential for conflict over natural resources increased after the Second World War when oil and natural gas displaced coal as the main source of energy for the industrialised countries. This happened because while most of the coal deposits were located in the industrialised countries and Australia, most of the accessible oil and gas reserves were located in the developing countries. This became apparent when the Central Intelligence Agency (CIA) engineered a coup against the democratically elected government of Mohammad Mosaddegh after it insisted upon auditing the books of the Anglo-Persian Oil Company and nationalised Iranian oil fields in 1953 when it refused to comply with is demand.

Another quantum jump in concern took place when three things happened in rapid succession between 1973 and 1976: OPEC quadrupled the price of oil; its Arab members imposed an oil embargo on the West in the wake of the Yom Kippur War and the US' dependence on imported oil rose to more than half of its needs. Another occurred between 2000 and 2005 when China's net imports of oil trebled from an average of 1.2 million barrels a day in 2000–2001 to 3.1 million barrels a day, 46 per cent of its total consumption, in 2005.[8] As a result, international oil prices jumped from $18 a barrel at the beginning of 2003 to $77 a barrel in the middle of 2006.

In 2009, although the industrialised nations consumed 53 per cent of the total oil output of the world,[9] they held only 5.5 per cent of its reserves.[10] Today, if we exclude shale gas, and oil trapped in tar sands, all but 6 per cent of the proven oil reserves that remain are located in the developing countries. This has set off a race to find or buy into new oil fields. An even greater need has been to ensure a smooth and uninterrupted flow of oil into the market. This has predisposed the industrialised countries to consider oil-exporting countries that have allowed the world market to dictate the rate of extraction of their oil or, better still, regulated its extraction to maintain a stable price in the oil market, as their allies, and those that have nationalised oil and gas reserves to tailor its extraction to their domestic needs, as their adversaries. In the Middle East, Saudi Arabia and the UAE have fallen into the former category, while Iran, Iraq and Libya have fallen into the latter.

The end of the era of cheap oil has heightened the potential for conflict. In 2011, Praveen Swami, then the diplomatic editor of *The Telegraph* pointed out:

> Five nations now account for more than half the world's proven reserves.... Saudi Arabia with 19.8 per cent, followed by Iran with 10.3 per cent, Iraq with 8.6 per cent, Kuwait with 7.6 per cent and the United Arab Emirates with 7.3 per cent. Oil producers outside the Middle East are running out of oil faster than

these five. Russia, which today produces almost as much oil as Saudi Arabia, will run out of oil by around 2020, the Institute for the Analysis of Global Security has estimated. Key African producers such as Nigeria, whose output today helps stabilise prices and meet significant parts of global demand, will also run dry by 2025 or so. That means that the Middle East will, in the not too distant future, be the world's principal source of oil—and there's no great imagination needed to see why betting on stability is probably unwise.[11]

Oil is not the only cause of conflict in the Middle East, but it is not entirely a coincidence that the United States and EU allied themselves with Saudi Arabia, the UAE and Kuwait, and against Libya, Iran and Syria, in the conflict that has been triggered by the Arab Spring. For while the first three are willing to let the 'world market,' i.e., the global oil majors, determine supply and prices, the latter three had insisted on keeping control of oil extraction firmly in their hands and tailor its sale to their budgetary needs.

As the world approaches the 'inflexion point' at which the annual addition to proven reserves falls behind the annual extraction of oil, the anxiety among oil importers over the security of supply from the Middle East, and therefore the potential for conflict, will continue to grow. The extraction of shale gas and 'tight oil' may give them a respite, but this is likely to be short-lived. A rapid shift into biofuels will allay this anxiety. Indeed had the shift to biofuels begun in the first decade after the oil price shocks of the 1970s, the world might have been spared much of the militarisation, conflict, death and destruction that it has witnessed in the past quarter of a century.

Thirty-five years after his project was shelved by MIT, Thomas Reed, who had begun experimenting with methanol as a transport fuel at MIT in 1973, speculated on his website, Woodgas:

It is interesting—and fruitless—to speculate on how history might have been different if the US had developed an aggressive synthetic fuel program in the 1970s. The threat of alternate production from gas, coal, or biomass would have tempered future rises in oil prices and reduced our funding of the Near East oil sheikhs. It would have prevented the Iran–Iraq war and the US–Iraq wars, and the terrorist attacks, and 9/11, and saved thousands of lives; it would have also saved millions of dollars. And we would have a much better estimate of the cost and means of producing the alternate fuels that we will need....[12]

To Reed's list, we must now add the secondary effects of the destruction of Saddam Hussein's regime in Iraq: heightened Israeli insecurity leading to attacks on Lebanon and Gaza; the destruction of secular and modern states in Libya and Yemen, the loss of more than 200,000 lives in Syria and, finally, the re-emergence of barbarism incarnate in the shape of ISIS and Al-Qaeda across the Arab world. Arresting climate change is a good reason to accelerate the shift out of fossil fuels, but helping to end this cycle of violence, and regression into savagery, is a better one.

Notes and references

[1] Strategic Foresight Group, 'Water Conservation for a Secure World: Focus on the Middle East' (Mumbai: Report by the Strategic Foresight Group), available at http://www.strategicforesight.com/publication_pdf/20795water-cooperature-sm.pdf (accessed on 28 October 2013).

[2] In meetings between Indian Prime Minister Man Mohan Singh and Premier Wen Jiabao at HuaHin, in Thailand, in 2009, and with President Xi Jinping at Durban, South Africa, in 2013.

[3] Lindsay Hughes, 'China's Plans for Yarlung–Tsangpo River Cause India Concern' (8 February 2013). Hughes is a Research Analyst for Indian Ocean Research Programme.

[4] In International law, first user rights start upon the completion of a project.

[5] For a detailed account of the reversal and the three years of rising tension along the LAC that followed, see Prem Shankar Jha in *India–China: Neighbours, Strangers*, edited by Ira Pandey (New Delhi: India International Centre).

6   J.S. Mill, *The Principles of Political Economy*, Book 4, Chapter 6: Of the Stationary State.

7   Praveen Swami, 'Why Oil Is So Important to China', *The Telegraph* (8 March 2011), available at http://www.telegraph.co.uk/finance/newsbysector/energy/oilandgas/8369641/Why-oil-is-so-important-to-China.html (accessed on 11 July 2017).

8   International Energy Agency, 'Oil and Gas Security Report 2011. Key Oil Data', available at https://www.iea.org/publications/freepublications/publication/China_2012.pdf (accessed on 11 July 2017).

9   US Energy Information Administration, 'International Energy Statistics, 2006–2010', available at http://www.eia.gov/cfapps/ipdbproject/iedindex3.cfm?tid=5&pid=54&aid=2 (accessed on 24 October 2013).

10   'BP Statistical Review June 2009', quoted in Wikipedia: Oil Reserves, available at http://en.wikipedia.org/wiki/Oil_reserves#Proven_reserves (accessed on 23 October 2013).

11   Praveen Swami, 'Why Oil Is So Important to China'.

12   Hammond's article and Reed's comment can be found at http://ourenergypolicy.org/docs/8/MethanolatMIT.pdf (accessed on 11 July 2017).

Index

About the Author

Prem Shankar Jha is an independent columnist, a prolific author and a former information adviser, Prime Minister's Office (India). He completed his master's in philosophy, politics and economics from the University of Oxford in 1961. After working for five years for the United Nations Development Programme in New York and Damascus, he returned to India to pursue a career in journalism. Over the past few decades, he has been Editor in *The Economic Times*, *The Financial Express*, and *Hindustan Times,* and the Economic Editor in *The Times of India*.

Jha has been a visiting scholar at Nuffield College, University of Oxford; Weatherhead Center for International Affairs and Fairbank Center for Chinese Studies, Harvard University; University of Virginia; University of Richmond and the Indian Institute of Management, Kolkata. He began research on alternate fuels in 1980. He has written extensively on them and their significance for the environment. In 1985, he was nominated to the energy panel of the World Commission on Environment and Development. In 1988, he was awarded the Energy Journalist of the Year award by the International Association for Energy Economics.

Jha has many published works to his credit. Some of them are given as follows: *India: A Political Economy of Stagnation* (1980), *In the Eye of the Cyclone: The Crisis in Indian Democracy* (1993), *A Jobless Future: Political Causes of Economic Crisis* (2002), *The Perilous Road to the Market: The Political Economy of Reform in Russia, India and China* (2002), *The Origins of a Dispute: Kashmir 1947* (2003), *The Twilight of the Nation State: Globalisation, Chaos and War* (2006), *Managed Chaos: The Fragility of the Chinese Miracle* (SAGE, 2009), *A Planet in Peril* (2009) and *Crouching Dragon, Hidden Tiger: Can China and India Dominate the West?* (2010).